MATERIAL SUBSTITUTION

MATERIAL SUBSTITUTION
Lessons from Tin-Using Industries

JOHN E. TILTON, Editor

RESOURCES FOR THE FUTURE / WASHINGTON, D.C.

Library of Congress Cataloging in Publication Data
Main entry under title:

Material substitution.

  Bibliography: p.
  Includes index.
  1. Tin. 2. Substitution (Technology) 3. Substitution
(Economics) I. Tilton, John E. II. Resources for the Future.
TS590.M37 1983   338.4   83-16164
ISBN 0-8018-3161-X

The cover was designed by Ruth Magann.

Distributed by the Johns Hopkins University Press, Baltimore, Maryland 21218
Manufactured in the United States of America.

Published November 1983

RESOURCES FOR THE FUTURE, INC.
1755 Massachusetts Avenue, N.W., Washington, D.C. 20036

Resources for the Future is a nonprofit organization for research and education in the development, conservation, and use of natural resources, including the quality of the environment. It was established in 1952 with the cooperation of the Ford Foundation. Grants for research are accepted from government and private sources only on the condition that RFF shall be solely responsible for the conduct of the research and free to make its results available to the public. Most of the work of Resources for the Future is carried out by its resident staff; part is supported by grants to universities and other nonprofit organizations. Unless otherwise stated, interpretations and conclusions in RFF publications are those of the authors; the organization takes responsibility for the selection of significant subjects for study, the competence of the researchers, and their freedom of inquiry.

This book was prepared in the Center for Energy Policy at Resources for the Future, Milton Russell, Director. It was edited by Ruth B. Haas and designed by Elsa B. Williams.

Contents

TABLES

FIGURES

Preface

Substitution following upon price increases has long been seen as the remedy for prospective materials shortages. However, substitution—its initiating causes and the process itself—has received little attention. Few studies describe in detail how one material comes to replace another in a specific product. Only generalities are known.

The study presented in this volume aims at injecting some realism into the topic. It describes in detail the role of tin in three applications and asks what produced the gradual displacement of the metal. As it turns out, the determinants of change vary over time and from case to case. Of particular importance is technological change, though government regulations, changing consumer tastes, and shifting reliability of supply are also among the motivating factors. Interestingly, the price of tin alone does not appear to provide either a direct or immediate impetus for change in the cases examined, though in the end, the new configuration, including the new material, must, of course, be economic to produce.

The study was funded by the National Science Foundation. Special thanks are due to Dr. Lynn Pollnow, under whose auspices the study was conducted. RFF became involved at the stage when most of the detailed work had been accomplished and the time had come to apply the findings more generally. Hans Landsberg organized a workshop aimed at exploiting information from other sources, generalizing the findings of the draft, and drawing policy inferences from them. The first chapter of this volume reflects the contribution of this workshop and the many helpful suggestions of the participants.

The workshop was supported by the National Science Foundation; subsequent work and the publication of this volume by Resources for the Future directly and by the RFF/Penn State Mineral Economics and Policy Program. This program, established in 1982, supports research on minerals problems, dissemination of research results, and the education of mineral economists.

July 1983

Milton Russell
Director, Center for Energy
Policy Research

Acknowledgments

A great number of people were helpful to the authors of this report. Lynn Pollnow of the National Science Foundation provided valuable guidance as grant officer and Simon Strauss of Asarco very kindly arranged introductions to a number of persons in tin-using industries.

It is impossible here to list all of those who provided time and information, but among them are Edward White, Associated Lead Industries; William T. Casey, Vinyl Additives; Mitchell S. Silkotch, Interstate Chemicals; Bernas A. Mitchell, Ford Motor Co.; Montfort Johnsen, Peterson/Puritan, Inc.; Donald Miner, Copper Development Association; Samuel Turkus, Bow Solder Products; Richard Brady, Federated Metals; H. L. Baker, W. J. Sanders, and L. P. Breuckel of the National Steel Corporation; W. A. Howell, U.S. Steel Corp., P. C. Katz, U.S. Brewers Association; J. J. Medich, Brockway Glass Co., R. W. Patton, Aluminum Company of America; M. E. Seehafer and M. Gregory, U.S. Steel Corp.; W. D. Way, Continental Can Co.; R. S. Weinberg, R. S. Weinberg and Associates; and R. Dennis, American Can Co.

It is no exaggeration to say that this book would never have seen the light of day without the long-suffering secretarial help of Kay Sefchick and Theonas Fleming of the Department of Mineral Economics at The Pennsylvania State University. They not only typed the dissertations that formed the basis of some of the chapters, but retyped and cleaned up several versions of this book while carrying a heavy work load. I am also very grateful to Ruth Haas at Resources for the Future for the care with which she edited the manuscript.

July 1983

John Tilton
Laxenburg, Austria

PART I

OVERVIEW

1

Material Substitution: Lessons from the Tin-Using Industries

John E. Tilton

Historically, the United States and other industrialized countries have substituted relatively abundant materials for increasingly scarce materials. This has alleviated—some would even say, postponed indefinitely—the upward pressure on material costs resulting from the depletion of high-grade, readily available, and easy to process mineral deposits.

Forest products, as Rosenberg (1973) has shown, provide a dramatic illustration from the past of this benefit of material substitution. The abundance of timber in nineteenth century America led to its widespread use, both as a fuel and as a building material. By the end of the century, however, the once seemingly endless tracts of forest land were gone, and many feared that shortages of this essential resource would soon curtail the country's economic development. The substitution of coal, petroleum, and natural gas for wood as a fuel, and iron and steel, aluminum, cement, and plastics for wood as a material, however, averted a shortage, and the country's economy continued to grow rapidly during the twentieth century. As for the future, Goeller and Weinberg (1976), Skinner (1976), and others contend that material substitution will have to play an even more critical role, if the adverse effects of dwindling mineral resources on economic growth and living standards are to be avoided.

While material substitution is an essential weapon in society's arsenal for holding the long-run problem of resource depletion at bay, it is important for other, more immediate reasons as well. Material substitution may, at least in some applications, soften the blow of abrupt and unexpected interruptions in mineral trade, whether these interruptions arise from wars, civil disturbances, strikes, natural disasters, or embargoes. During World War II, Germany, the United States, and other countries managed to replace tin, tungsten, and other restricted materials with more available alternatives in many end uses (Eckes, 1979, chap. 4). In recent years, interest in material substitution as a cushion against supply interruptions has grown as the political situation in southern Africa and other mineral producing regions has become more turbulent.

Despite its importance, we still have much to learn about the nature of material substitution. For example, just how important a factor is substitution in shaping short-run and long-run trends in material consumption? Are its effects generally evolutionary and continuous, or abrupt and discontinuous? Is the replacement of one material by another primarily motivated by shifts in relative prices, or are changes in government regulations, consumer preferences, technology, and other factors typically more influential? Where substitution does occur in response to shifts in material prices, what is the nature of the time lag between the two? What kinds of substitutions can be made quickly, and what kinds require years or decades to accomplish?

Approach

To provide insights into such questions, material substitution was examined in three tin-using industries—beverage containers, solder, and tin chemical stabilizers used in the production of plastic pipe. In these case studies, which are found in part II, material substitution is defined to encompass a number of different types of events that may substantially alter material use.

The first is material-for-material substitution, where one material is used in place of another. Examples include the use of an aluminum beer can for a glass bottle, the installation of plastic rather than copper pipe, and using aluminum rather than copper-brass radiators to cool automobile engines.

The second is other-factors-for-material substitution, where consumption is reduced by increasing nonmaterial inputs, such as labor, capital, and energy. The hand soldering of radios, televisions, and other electronic products, for example, requires less solder than more automated production using printed circuit boards. As indicated in part II, however, producers have favored printed circuit boards despite their higher material intensity because they lower labor costs.

The third is quality-for-material substitution, where material is saved by reducing the quality or performance standards of the final product. For example, the lightweight, nonreturnable bottles produced during World War II had relatively high breakage rates. These bottles could have been strengthened by using more glass to make them thicker.

The fourth is interproduct substitution, whereby a change in the mix of goods used to satisfy a given need alters the demand for one or more materials. Television may reduce the need for movie houses, public transportation for private automobiles, and the telephone for letters or visits. This type of substitution, unlike others, does not affect the manufacturing process or the materials used in the production of individual goods, but instead influences material use by changing the composition of the goods and services produced. It includes what Chynoweth (1976) and others call functional substitution, which occurs when a product (along with the materials it embodies) is replaced by an entirely different means of achieving the same end. The use of satellites instead of underground cables to transmit long-distance telephone messages is an example.

The fifth is technological substitution, in which an advance in technology allows a product to be made with less material. The introduction of electrolytic tinplating during and after World War II greatly reduced the tin requirements for beer and soft drink cans, and provides a good example of this type of substitution. New technology can also create new products, as the aluminum can illustrates, and thus increase the opportunities for interproduct substitution.

In short, material substitution may result from the introduction of new technology, from shifts in the composition or quality of final goods, and from changes in the mix of factor inputs used in producing these goods. This is clearly a broader and more encompassing definition than the common perception of material substitution, which often is restricted solely to material-for-material substitution. A broad definition, however, is more appropriate and useful if ultimately one is concerned with alleviating material shortages or forecasting future mineral requirements.

There are, of course, other ways to study material substitution than the case study approach followed here. One possibility entails the estimation of production or cost func-

tions for entire industries, economic sectors, or even the economy as a whole. Such efforts, which Slade (1981) has recently reviewed, generally treat materials as a single factor of production, and assess the extent to which they can be replaced by other inputs such as capital, energy, and labor. Their findings vary greatly, depending on how the production or cost functions are specified, how technological change is treated, and how other issues are resolved. They also tend to suffer from the use of aggregate input data and from their static nature (Kopp and Smith, 1980; Slade, 1981).

A second alternative involves more micro investigations that consider individual materials and often focus on specific end uses. These efforts specify and estimate formal models, the most common of which are econometric supply and demand models, such as the Fisher, Cootner, and Baily model of copper (1972) and the Woods and Burrows model of aluminum (1980).

These models typically assess the effects of material substitution by including in their demand equations price variables for close substitutes, and so implicitly assume that material prices are the primary driving force behind substitution. The specification of such models also normally presumes that the relationship between demand and material prices adheres to a particular lagged structure that remains fixed over time, and that this relationship is continuous and reversible. The latter implies that the demand lost when a material's price increases is recovered when price returns to its original level. While they are perhaps plausible in certain instances, such assumptions are open to question. Even more important, in making them, these models are assuming away many of the interesting questions associated with material substitution.

The decision to concentrate on a limited number of in-depth case studies was taken in the belief that this was the most promising approach given the current state of understanding and knowledge about the material substitution process. Like other approaches, it has its limitations. In particular, since material substitution occurs constantly in literally thousands of products throughout the economy, only tentative conclusions about the general nature of material substitution can be drawn on the basis of such a limited sample of actual situations.

The focus on tin-using industries was motivated by two considerations. First, the price of tin rose substantially during the 1970s, both in real terms and in relation to the prices of most of the alternative materials with which it competes. In this regard, there is some concern that the depletion of known, high-quality deposits and the failure to discover new deposits of comparable quality will continue to force prices up in the future. Second, the United States and other industrialized countries rely almost entirely on imports from developing countries for their primary tin requirements. These imports come in large part from Malaysia and other countries in Southeast Asia. The lengthy sea routes over which they travel could easily be interrupted during military or other emergencies.

While all of the case studies examine uses of tin, they are not concerned exclusively with tin. Indeed, by its very nature one cannot study material substitution without considering more than one material. In its many applications, tin competes with a variety of materials, including chromium, steel, glass, aluminum, cast iron, copper, antimony, lead, plastic, and silver. In certain applications steel, plastics, lead, and other materials are complementary products in the sense that they are used with tin in producing a product. As a result, a reduction of their price tends to increase the demand for tin.

The three sectors of the tin industry selected for study—beverage containers, solder, and tin chemicals used in the manufacture of plastic pipe—were chosen for different reasons. As every consumer of beer or soft drinks knows, the beverage container market has been a lively battlefield for competing materials over the last two decades. The sturdy returnable glass bottle, the lighter one-way bottle, the tinplate can, the aluminum can, the tin-free steel can, and recently the plastic bottle have all fought for this market. The variety of competing materials and the dramatic speed with which their fortunes may rise and fall make the container market of particular interest.

In contrast, it is widely assumed that little material substitution has occurred in the solder market, and that little is possible. Indeed, solder by definition is an alloy of tin and lead, with other materials added in minor amounts in certain applications. Moreover, for technical reasons the tin content of solder cannot be varied without seriously affecting its performance in certain uses, particularly electronic equipment. So solder provides an opportunity to examine the extent and nature of material substitution in applications where substitution is generally presumed difficult. As solder is consumed in a multitude of products, this case study concentrates primarily on its use in can seams, new motor vehicle radiators, and fillers for automobile bodies. Other important applications of solder in the electronics and plumbing industries are considered in less detail.

Tin chemicals, and particularly organotin chemicals used as stabilizers in the production of plastic pipe, have grown substantially over the last two decades. In contrast, as table 1-1 shows, most sectors of the tin industry have been stagnant or decreasing. Consequently, an investigation of tin chemicals provides an opportunity to examine why material substitution may not always reduce the demand for a commodity whose price is rising rapidly. Due to the many uses of tin chemicals, the focus is restricted to organotin chemicals and their use in plastic pipe production.

The scope of the case studies is further limited in three respects. First, although many of the trends in U.S. tin consumption are also found in other countries, in a number of instances important differences do exist between the United

Table 1-1. Tin Consumption in the United States by End Use Sector, 1955 and 1978

End Use Sector	1955 Thousands of Tons[a]	1955 Percent	1978 Thousands of Tons[a]	1978 Percent
Tinplate	34.1	37	17.3	27
Solder	22.6	25	18.3	29
Bronze and Brass	20.0	22	10.4	17
Chemicals[b]	1.9	2	7.6	12
Other[c]	13.3	14	9.5	15
Total	91.9	100	63.1	100

Source: American Bureau of Metal Statistics (various years).
[a]Metric tons are used throughout this study.
[b]Includes tin oxides and miscellaneous.
[c]Includes terne plate, babbit, collapsible tubes, tinning, pipe and tubing, type metal, bar tin, miscellaneous alloys, and white metal.

States and other countries. While some of these differences are noted in passing, the emphasis is on material substitution in the United States.

Second, the case studies do not cover all the important tin consuming industries. Table 1-1, which shows the major end use sectors for tin in the United States for the years 1955 and 1978, indicates that over two-thirds of the country's supplies are used to make tinplate, solder, and bronze and brass. Beverage containers are part of the tinplate sector, but so are fruit and vegetable, meat, soup, and other cans. Beer and soft drink cans have accounted for only between 10 and 20 percent of total U.S. tinplate consumption. The case studies also examined important uses of solder and chemicals, but as noted already, coverage of these sectors is far from complete.

Third, the studies concentrate on explaining the past, primarily the period since 1950, rather than predicting the future. The past can be documented with data, and the factors causing material substitution identified more or less. The future is far more uncertain, and detailed information on material consumption, prices, and other relevant variables is not available. Thus, the analyses carried out here are designed to enhance knowledge of the nature and determinants of material substitution by improving our understanding of events that have taken place. Of course, it is hoped that this information will ultimately prove useful in assessing the role that material substitution may play in the future.

The case studies are similar in that they all follow three analytical steps. In the first, the pounds or tons of tin consumed in specific applications, such as soft drink containers, are quantified for the years covered in the analysis.[1]

In the second step, what are called the apparent determinants of tin consumption are identified and measured. These determinants are directly related to tin consumption by an identity, and basically indicate how tin usage has

___
[1]Metric tons are used throughout this study.

changed over time. For example, the amount of tin in the solder used to produce new motor vehicle radiators has varied over time as a result of changes in (1) the number of new motor vehicles and hence radiators produced; (2) the proportion of radiators made from copper and brass, rather than aluminum, and so requiring soldering; (3) the pounds of solder used per copper-brass radiator; and (4) the tin content of the solder used for this purpose. Any change in tin consumption from one year to another in radiator solder must result from a change in one or more of these apparent determinants.

Although the apparent determinants vary from one end use to another, they contain in all instances one variable that reflects the change over time in the output of the end use. In the preceding example, it is the number of new motor vehicle radiators. This apparent determinant may change over time for reasons other than material substitution. The number of new motor vehicles and in turn radiators produced in the United States, for instance, may increase simply because population and per capita income are growing. Changes in all the other apparent determinants, however, are the result of one or more of the five types of material substitution identified earlier. Moreover, while the apparent determinant reflecting changes in the output of the end product can change for other reasons, it too may be affected by material substitution, particularly interproduct substitution. The growth of mass transit systems in metropolitan areas, for example, may slow the growth of new automobile radiators.

By identifying the apparent determinants and empirically assessing their effects, one can dissect the change in tin consumption into its component parts. The reasons why these parts change can then be assessed.

This is done in the third step, which identifies and evaluates the major underlying factors responsible for the changes in the apparent determinants, and thus ultimately in tin consumption. These factors include the price of tin, the price of alternative materials, technological developments, government regulations, and a host of other considerations. At this step, the analysis cannot be as empirically rigorous. Assessing the relative importance of the major underlying factors involves some judgment, even after weighing the available information from industry and other sources. Surprisingly, however, in many instances, due largely to the level of disaggregation on which the analyses are conducted, the important underlying factors are readily apparent.

Findings

In all of the end uses studied—beer and soft drink containers, can seams, motor vehicle radiators, automobile body solder, and chemical stabilizers used in the production of the PVC plastic pipe—material substitution greatly affected tin consumption over the longer run, a period of ten years or more. Moreover, in many instances, it sharply altered or reversed trends in tin usage even in the short run.

The tinplate can, for example, after years of increasing its share of the beer container market, abruptly found itself during the late 1960s being pushed out of this market by aluminum and tin-free steel cans. Even with solder, often considered immune to material substitution, the introduction of low-tin alloys, produced by substituting lead and minute amounts of other metals for tin, substantially reduced the need for the latter in can seams and automobile body solder. More recently, the trend away from large automobiles has decreased the size of the average motor vehicle radiator and in turn the need for solder and tin in this use.

Two types of material substitution are especially prevalent. The first is material-for-material substitution. The beverage container market in particular has experienced substantial changes over time in the number and quantity of materials it consumes. A wide variety of materials also compete for the pipe market. Indeed, much of the growth in tin chemicals for this industry has come about as a result of the substitution of plastic for copper, cast iron, steel and other kinds of pipe. In solder, material-for-material substitution has occurred on a modest scale with the appearance of aluminum radiators, and on a more significant scale with the widespread adoption of low-tin alloys for can seam and body solder.

Technological substitution is the second type often encountered. New technology reduced the tin content of the average size tinplate beverage container by 93 percent between 1950 and 1977. Similarly, in the plastic pipe industry, the introduction of second and third generation organotin stabilizers caused their tin content to fall by nearly 50 percent.

Other types of material substitution, though apparently less common, also occur. The growing distribution of soft drinks in bulk containers, due in large part to the rise in fast food chains, slowed the penetration of the tinplate can in this market. Another example of functional substitution is found in the electronic industry where the miniaturization of components greatly reduced the number of electrical connections requiring soldering.

The case studies also strongly suggest that intermaterial competition and substitution are becoming more intense and prevalent over time, as the number of materials increases and the properties of existing materials are enhanced, allowing them to penetrate new markets. This tendency is clearly found in the beverage container market. In the early 1950s, the glass bottle basically monopolized both the packaged beer and soft drink markets. The only competition came from the tinplate can, in which a modest amount of beer was shipped. Starting in the 1960s and continuing in the 1970s, however, the bottle encountered increasing competition from the tinplate can, the aluminum can, the tin-free steel can, and recently the plastic bottle. While the glass bottle continues to be the most popular packaged container (if the returnable and one-way bottle are considered together), its share of the soft drink market dropped from 100 to 62 percent between 1950 and 1977. In beer, it fared even less well, retaining only 40 percent of the market by 1977.

Growing material competition is found in other sectors as well. Polyvinyl chloride (PVC) plastic did not enter any of the pipe markets examined until the early or mid-1960s. Antimony stabilizers as an alternative for organotin first appeared in the late 1970s. Solder, once required to seal the side seams of all cans, had its monopoly of this market broken by the development of welded and cemented seams, and by the introduction of the two-piece can, which has no side seam.

The factors responsible for material substitution are numerous and their significance tends to vary over time and by end use. However, three factors—relative material prices, technological change, and government regulations—are of particular importance in all of the case studies. In examining their influence, it is useful to start with material prices, since economists and others often assume that they are the principal motivation or incentive for material substitution. On this, the evidence is mixed; or more correctly, it supports the proposition that material prices are important but that several qualifications or corollaries to this conclusion are necessary.

There is no doubt that the high and rising price of tin has discouraged its use. As table 1-1 indicates, total tin consumption in the United States, in contrast to that of nearly all other metals, has declined over the postwar period. Moreover, it is not difficult to document specific examples of substitution away from tin that is motivated at least in part by its price, as is illustrated by the development of second and third generation tin stabilizers, low-tin alloy solders for can seams and automobile body fill, and the declining tin content of tinplate.

Still, material substitution has in many instances increased, and increased substantially, the use of tin. Such substitutes take place, despite the high price of tin, for two reasons. First, in some uses tin provides superior quality or performance that outweighs its higher cost. For example, electronic manufacturers, after experimenting with solders containing 50 percent tin and 50 percent lead, reverted back to 63 percent tin solders because the latter's lower melting temperature resulted in less damage to printed circuit boards during the manufacturing process.

Second, substitution takes place on many levels. In the pipe industry, for example, it occurs in the production of stabilizers, plastic compounds, and pipe.[2] Since tin consti-

tutes between 18 and 35 percent of the final cost of producing organotin stabilizer, producers are strongly motivated to reduce or eliminate their use of tin. However, at subsequent stages of production, tin becomes an increasingly smaller fraction of total cost, finally accounting for 1 percent or less of the final price of PVC plastic pipe. At this stage, the price of tin has much less of an effect on the type of pipe purchased.

So, it is not surprising that the price of tin has its greatest impact on material substitution at relatively early stages of production, where tin costs constitute a significant portion of total costs. It is here that substitution has with considerable consistency reduced tin usage. In plastic pipe, for example, the tin content of stabilizers has dropped substantially as a result of the development of second and third generation tin stabilizers and the recent appearance of antimony stabilizers. This contrasts with the material substitution at later stages of production, which has significantly increased the use of tin. At the compound stage, polyvinyl chloride, which requires a stabilizer, has replaced other types of plastic; and at the pipe stage, plastic has replaced pipe made of copper, cast iron, and other materials. At both of these stages, the price of tin has had a small impact on final cost, and has been easily offset by other considerations.

When tin prices do stimulate material substitution, three possible time dimensions or lagged responses are found. First, where existing technology and equipment permit the use of one material for another in the production process, substitution can respond immediately to changes in material prices. In such situations, when the price of a material rises above a certain threshold level, producers switch away from it. When the price falls below this threshold, they switch back again. Not only is the response fast, but it is reasonably predictable. The case studies of containers, solder, and tin chemical stabilizers uncovered a few instances of this type of response. The dual canmaking line, introduced about 1976, can substitute between aluminum sheet and tinplate in 4 hours. Also, plastic pipe producers can switch from one plastic resin to another quite quickly since the same equipment is used. However, the opportunities for such a rapid response to changes in material prices or other conditions are limited.

Second, where the technology for substitution exists but equipment must be altered or completely replaced, material substitution occurs only after some delay. Moreover, the lag normally exceeds the minimum period required to build new facilities or modify old ones, for producers hesitate to make such changes, in light of the costs, until they are

---

[2]The production of plastic pipe involves the melting and extruding of a plastic compound, which is composed of a plastic resin, such as polyvinyl chloride (PVC) or acrylonitrile-butadiene-styrene (ABS), plus several chemical additives. The latter include lubricants to prevent the compound from sticking to the extruder, impact modifiers to increase resilience, and

flame retardants. In addition, compounds made from PVC resins must contain a stabilizer to prevent decomposition and other undesirable effects caused by heat during production. Organic tin chemicals or organotins are the most commonly used stabilizers, though lead and antimony stabilizers are also available. The stages of plastic pipe production are described in more detail in part II.

reasonably certain that the change in material prices is not temporary. The expense of switching also means that the threshold price at which substitution occurs is higher than would otherwise be the case, and that once a change is made, the price of the replaced material must fall appreciably below the original threshold level before producers will switch back again. This type of substitution has occurred from time to time in the beverage container and pipe markets. Though not all that prevalent, it is more frequently encountered than the first type of response.

Third, where new technology must be introduced before substitution occurs, the response to a change in material prices takes even longer. In this case, there is also greater uncertainty and variation regarding both the size of the response and the length of its time lag. Once substitution away from a material does occur, a decline in its price is even less likely to result in the recapturing of a lost market than in the second situation.

It is this third type of response that has consistently had the greatest impact on tin usage. In beverage containers, for instance, tin prices have encouraged the introduction of electrolytic plating, differential and lighter tin coatings, the tin-free steel can, and the tin-free steel bottom used on all tinplate cans—all major developments reducing tin usage. In solder, the experiments that led to the use of low-tin alloys for can seams and automobile body finishing have been particularly important. In the pipe industry, tin consumption has been cut by about 50 percent from what it would otherwise have been by the development of superior tin stabilizers and the introduction of antimony stabilizers. Thus, the relatively high price of tin, when it has effected material substitution, has often done so indirectly by stimulating new tin-conserving technologies.

This conclusion is part of a broader finding. In all of the end uses examined, technological change appears as the dominant factor affecting material substitution. The use of tin in beer and soft drink containers, for example, was made possible by the development of the tinplate can. Its competitiveness over time has been enhanced by electrolytic plating, the easy-open aluminum top, the cheaper tin-free steel bottom, two-piece production techniques, and a host of other innovations. Conversely, the development of the one-way bottle, the aluminum can, the tin-free steel can, and subsequent improvements in these containers have reduced tin consumption in this market.

The use of solder in electronics was first stimulated by the introduction of printed circuit boards, and then reduced by the trend toward miniaturization. In cans, it declined substantially after welded and cemented seams were developed and two-piece canmaking technology was introduced. In motor vehicle radiators, the appearance of the sweat-soldered tube and corregated fin core for the dip-soldered cellular core used in the early postwar period substantially reduced solder and in turn tin usage.

The use of tin as a stabilizer for the production of PVC plastic was originally made possible by new technology. The rapid penetration of PVC in the pipe market that followed was stimulated by advances in extruder techniques, the development of superior tin stabilizers, and other innovations. After this rapid ascent, tin consumption in PVC plastic pipe may decline as quickly as it rose due to another new development, antimony stabilizers.

These are but a few of the many developments identified in part II. Technological change has been a powerful force shaping the use of tin in all of the applications examined. Moreover, its impact has often been abrupt and uneven, at times stimulating and at other times curtailing tin use. This random and discontinuous character of technological change makes it difficult to foresee its effects.

Government regulations have also influenced material substitution. While less important than technological change, the influence of this factor appears to be growing over time. Recent container deposit legislation, for example, has favored the returnable glass bottle; and among metal containers, the aluminum can, whose homogeneous composition makes it cheaper to recycle. During military emergencies the government has restricted the use of tin containers in the domestic market, and encouraged tin-conserving new technologies, such as electrolytic plating and low-tin solder alloys. Tin consumption in solder for cans has been influenced by the regulations of the Food and Drug Administration, in automobile radiators by the fuel efficiency standards of the Environmental Protection Agency, and automobile body finishing by the lead exposure standards of the Occupational Safety and Health Administration. In the pipe industry, the ubiquitous building codes imposed by thousands of local authorities have encouraged the use of tin by inhibiting lead stabilizers and slowing the growth of antimony stabilizers. They have also impeded tin consumption by delaying the use of plastic pipe.

While government regulations, material prices, and particularly technological change are the more prevalent and influential factors responsible for material substitution, other considerations have also been important in certain applications. For example, changes in customs, causing the decline of the local tavern and the rise of fast food chains, have altered the mix of packaged and bulk containers in the beer and soft drink markets. The popularity of the vinyl-roofed automobile has reduced the consumption of body solder, while the desire for quieter plumbing systems has favored heavier materials over plastic.

Implications

The findings described in the preceding section point toward several general conclusions regarding the nature of material substitution. First, such substitution may, and often does, substantially alter material requirements. This is particularly true over the long run, but applies at times in the short run as well. Second, material-for-material and technological substitution are the easiest types of substitution to identify and appear more prevalent than other types. Third, a major cause, perhaps the major cause, of substitution is technological change. Fourth, changes in material prices typically have little effect on the mix of materials in the short run because producers are constrained by existing technology and equipment. Over the long run, they have more of an impact. Indeed, the major influence of material prices apparently is exerted indirectly over the long run by altering the incentives to conduct research and development on new material-saving technologies. Finally, government activities and regulations often motivate material substitution.

These conclusions, of course, are tentative. As pointed out earlier, material substitution is taking place continuously in a multitude of products throughout the economy, and so there are dangers in drawing general conclusions on the basis of only a few cases. It is important to keep this caveat in mind in examining the implications of the findings.

The Demand Curve

The downward sloping demand curve, a conceptual tool widely used by economists, business analysts, and others, indicates the quantities of a commodity that the market will demand at various prices over a particular period, such as a year, on the assumption that all other determinants of demand remain unchanged. Although the curve is formally derived in microeconomics from the theories of the firm and consumer behavior, the nature of the curve and in particular its downward slope seem reasonable. As the price of a good goes up, less will be demanded, first because consumers will substitute other goods whose prices have not risen (the substitution effect) and second because consumers will have less real income to spend on all goods and services (the income effect).

Materials are rarely desired for their own sake, but rather their demand is derived from the demand for final goods and services. Moreover, the proportion of total costs contributed by any particular material in the production of most finished products is small. This means that changes in material pieces generally do not produce major shifts in the output of final goods and services. Nor do they cause significant changes in real consumer income. As a consequence, the reduction in material demand resulting from an increase in price comes about entirely, or almost entirely, as a result of material substitution.

As commonly drawn, the demand curve presumes that the functional relationship between price and demand is reversible. Frequently, however, commodity analysts claim, if a material loses a particular market, that market will be lost forever. Such statements imply that an industry may

not be able to recapture a market lost during a price rise even if its price subsequently returns to its previous level. In other words, if a commodity moves up its downward sloping demand curve, it may not be able to reverse itself and move back down the same curve, as the conventional demand curve implies. Such concerns, it is even suggested, help explain the restraint that molybdenum, nickel, aluminum, and other material producers exercise in raising their prices during boom conditions.[3] As is well known, many material producers charge less than the market clearing price at such times.

In assessing the reversibility assumption, it is useful to distinguish among short-run, medium-run, and long-run demand curves. The short-run demand curve indicates how demand responds to a change in price during a time period that permits neither plant and equipment nor technology to change. The medium run covers a period sufficiently long to allow plant and equipment to change, and the long-run demand curve for technology as well as plant and equipment to change.

The assumption of reversibility seems most plausible for the short-run demand curve. The material substitution that can take place in the short run, such as the use of aluminum sheet for tinplate on a dual canmaking line, involves changes that can be made quickly with little cost or disruption. When these conditions govern the switch from one material to another, they are also likely to hold for a switch back to the original material. While the number of such substitution opportunities is limited, causing the short-run demand curve to be relatively steep, the possibilities that do exist and that can be quickly exploited tend to be easily reversible.

Reversibility appears more questionable over the medium run. Material substitution is in such instances likely to entail considerable expense. New equipment may have to be ordered, personnel retrained, and production lost during the changeover period. Once a firm has incurred such conversion costs, it will not find it worthwhile to switch back to a material unless the latter's price drops considerably below the level at which the original substitution became attractive.

In the long run, the assumption of reversibility appears particularly doubtful. Within this time frame, price-induced innovations may substantially alter the underlying technical and economic conditions governing the use of a material in a number of its applications. The introduction of better

---

[3] The argument that producers keep prices below what the market will bear during economic booms to prevent substitution is not by itself very convincing. Since demand exceeds supply when price is held below the market-clearing level, producers must allocate or ration their customers. This means that once a firm has received its quota, the price of additional supplies is infinitely high. This presumably should encourage customers as much or more to search for possible alternative materials. The argument becomes more plausible if producers while restraining prices also discriminate in their rationing in favor of those customers with the greatest substitution possibilities. The extent to which producers actually engage in such discrimination, however, is unknown. Normally, it appears they allocate supplies on the basis of their customers' past purchases.

products can prevent a resurgence of consumer demand for older, traditional products even when the raw material costs of the latter fall sharply. Who, for example, would return to mechanical calculators even if the cost of their embodied materials were zero? Such uncertainty surrounding the generation of new technology makes it most unlikely that the demand induced by a fall in price will exactly offset the demand lost by an equivalent rise in price.

The demand curve is also often assumed to be continuous. The importance and nature of material substitution, however, suggest that this too may not be very realistic, particularly for those materials whose use is concentrated in a few major applications. Price may rise over a range, with little or no effect on demand. Then, once a particular threshold is crossed, making the use of an alternative material attractive, demand may fall sharply. Such discrete jumps may be found in both short- and medium-run demand curves. They are even more likely to distinguish the long-run demand curve, since innovations, such as the antimony stabilizer or the aluminum can, by their nature are discrete events that either do or do not occur. When they do occur, their impact can be substantial.

A third questionable but common assumption concerns the stability of the relationship between demand and price over time. While the demand curve is often assumed to shift in response to changes in income, the prices of substitute materials, and other factors affecting demand, once the influence of these variables is taken into account or controlled for, the relationship between the price of a material and its demand is presumed to be reasonably stable from one period to the next. Efforts to estimate demand curves, or entire demand functions, on the basis of time series data must make such an assumption, at least implicitly.[4] The use of demand functions, however estimated, for forecasting or assessing future market conditions also requires this assumption.

The importance of price-induced technological change, however, suggests that such stability may not exist. The occurrence of innovations inherently involves a certain random or chance element. In addition, their effect on demand varies greatly. In short, there is not a stable relationship between price and the number of induced innovations, or between the number of induced innovations and their cumulative effect on demand. Yet both of these conditions are needed if the response of demand to a change in price is to remain stable over time.

Since the demand curve is defined as the relationship between the demand for a commodity and its price, assuming all other factors remain unchanged, one way of trying

to preserve the assumptions of reversibility, continuity, and intertemporal stability is simply to exclude the effects of price-induced technological change. By definition, one might argue, such developments involve a change in one of the other factors (technology) which the construction of a demand curve can presume remains constant. This approach is certainly feasible, and it increases the plausibility of the three assumptions. Unfortunately, if the principal effect of a change in price on material demand occurs indirectly via induced technological change, as the studies here suggest, this approach reduces the demand curve to a sterile academic concept with little practical use. Indeed, by ignoring the major impact of price, it may even be misleading.

Material Shortages

In assessing the role that material substitution can play in mitigating or eliminating shortages, two very different types of shortages must be distinguished. The first is due to the depletion of high quality mineral deposits, and imposes upon society the necessity of procuring its mineral needs from increasingly costly sources. This type of shortage comes about slowly with considerable warning, as the real price of a material climbs persistently over time and in the process gradually constricts demand. Although examples are hard to find because the real costs of most materials have actually fallen over the last century, there is concern that this type of shortage may become a serious problem some time in the future.

The second type of shortage, in contrast, is quite common. It tends to be temporary, rarely lasting for more than three to five years, and often arises quickly with little warning. It can be caused by a variety of factors, including war, embargoes, cyclical surges in demand, strikes, accidents, natural disasters, and inadequate investment in mining and processing. It may manifest itself in the form of sharply higher real prices, or where a few major firms maintain a relatively stable producer price, in the form of actual physical shortages.

The case studies suggest that material substitution can make a major contribution toward alleviating the first type of shortage, but much less of a contribution toward the second type. Substitution has substantially reduced tin consumption in a number of end uses, and the high and rising price of tin appears to have been a significant factor, at least indirectly, in motivating this shift. However, the response to changing material prices is typically small in the short run, for rarely is one material substituted for another while the same equipment, the same technology, and the same production processes continue in use. At a minimum, material substitution normally involves at least the replacement or modification of existing equipment, and more commonly, the development of new technologies. Since such changes take time, the major impact of a price rise on the con-

---

[4]The demand function indicates the relationship between demand and all factors influencing demand. It is thus more encompassing than the demand curve which portrays the relationship between a commodity's price and its demand, assuming that income and other factors affecting demand remain constant at some prescribed level.

sumption pattern of a material is not realized for a number of years. By then, the second type of shortage is usually over.

Market Power and Producer Cartels

In assessing the market power of the major material producers, most research has emphasized market concentration, which reflects the number and size distribution of the firms in an industry. Since many metal and other material industries are fairly concentrated, this approach often leads to the conclusion that firms do possess market power and the ability to earn excess profits. Moreover, in these oligopolistic industries the major producers frequently quote or set a producer price, rather than simply accept a price determined on a competitive exchange. This behavior is often cited as collaborating evidence for the conclusion that these firms possess market power.

During the 1970s, concern over the possible exercise of monopoly power expanded to encompass the governments of mineral producing countries, as well as the multinational mining corporations and other major producing firms. The success of OPEC in raising the price of oil in 1973 led many to conclude that other mineral-producing countries would also attempt to form producer cartels and artificially raise material prices.

Material substitution, however, can severely limit market power even in highly concentrated industries. Collusive efforts by established producers to raise prices substantially are likely to stimulate new technological activity, and eventually end in failure, with markets irretrievably lost. While higher prices may have little effect on material substitution in the short run, permitting a cartel to succeed for a while, the adverse consequences over the longer term are likely to far outweigh the short-term benefits. This discourages collusive activities except where firm managers or government officials have unusually short time horizons or are ignorant of the long-term consequences.

This conclusion, coupled with the findings that intermaterial competition is growing more intense over time as new materials are developed and traditional ones expand into new markets, suggests that the conventional view of market power in the material industries, based primarily on considerations of market concentration and pricing behavior, may be twenty years out of date. It also suggests that Schumpeter (1950, pp. 84–85), writing in the early 1940s, correctly described the nature of competition in the material industries today, if not then:[5]

. . . It is still competition within a rigid pattern of invariant conditions, methods of production and forms of industrial organization in particular, that practically monopolizes attention. But in capitalist reality as distinguished from its textbook picture, it is not

[5]Reproduced with permission from J. E. Schumpeter, *Capitalism, Socialism, and Democracy,* Harper and Row, New York, 1950.

that kind of competition which counts but the competition from the new commodity, the new technology, the new source of supply, the new type of organization (the largest-scale unit of control for instance)—competition which commands a decisive cost or quality advantage and which strikes not at the margins of the profits and the outputs of the existing firms but at their foundations and their very lives. This kind of competition is as much more effective than the other as a bombardment is in comparison with forcing a door, and so much more important that it becomes a matter of comparative indifference whether competition in the ordinary sense functions more or less promptly; the powerful lever that in the long run expands output and brings down prices is in any case made of other stuff.

It is hardly necessary to point out that competition of the kind we now have in mind acts not only when in being but also when it is merely an ever-present threat. It disciplines before it attacks. The businessman feels himself to be in a competitive situation even if he is alone in his field or if, though not alone, he holds a position such that investigating government experts fail to see any effective competition between him and any other firms in the same or a neighboring field and in consequence conclude that his talk, under examination, about his competitive sorrows is all make-believe. In many cases, though not in all, this will in the long run enforce behavior very similar to the perfectly competitive pattern.

Forecasting Material Requirements

Government agencies, private firms, international organizations, independent consulting firms, and others forecast future material requirements. These forecasts are needed to formulate sound public policies in the resource field, as well as for private investment decisions.

Although many different types of forecasting procedures are used, these techniques can be separated into three generic groups. The first encompasses a variety of statistical procedures that differ greatly in their complexity, but basically involve analyzing past trends and projecting them into the future. One such technique is simply extending into the future the linear trend of past consumption. Far more complicated is the Box-Jenkins procedure, which employs a mathematical model to identify various functional trends present in past consumption and to estimate the parameters of these trends. It then uses these estimated trend functions to project consumption into the future.

The second group of forecasting methods is comprised of models that specify causal or behavioral relationships. Traditional supply and demand analyses fall into this category. In these models, demand is typically assumed to be a function of the price of the commodity, an income or activity variable, and perhaps the prices of complementary and substitute products as well. Since the full effect of a change in price on demand may not occur immediately, various types of lag responses may be built into such functions. Once the causal equations are identified, their parameters must be estimated. This is usually done on the basis of past data and behavior using econometric techniques, though parameters may be determined on the basis of known technical relationships or other a priori information.

The third group includes forecasting techniques that are qualitative or judgmental in nature, and less quantitative. The Delphi method in which the views of a number of experts are solicited and then integrated to produce a forecast falls into this category. So too do forecasts based on the informed judgment of analysts who have adjusted current trends in consumption or the predictions of causal models for the effects of anticipated changes in public policies, consumer preferences and habits, technology, and other factors whose future influence on consumption cannot be measured with precision.

Material substitution greatly complicates the task of forecasting mineral requirements, and is likely to render any technique regardless of the category to which it belongs vulnerable to wide margins of error when forecasting over the longer term, ten to twenty years into the future. In the short run, the effects of material substitution may be small or relatively continuous, and so more predictable. Yet even this is not certain. Within a year the strong upward market penetration of a product can be accelerated or completely reversed by material substitution. Moreover, the growing intensity of material competition over time makes this type of volatility increasingly likely.

The first group of forecasting techniques implicitly assumes that material substitution is either a relatively unimportant factor shaping consumption, or that it is continuous and evolutionary so that its future influence can be projected from its past effects. Neither of these assumptions is reasonable over an extended period of time, and as just noted, may not even be valid in the short run.

The second group of forecasting techniques, when it considers material substitution explicitly, generally assumes that it is motivated or caused by a shift in relative material prices. This is the case, for example, when the demand functions within a supply and demand model include price variables for major substitute commodities. Moreover, although such models do not necessarily assume that a change in material prices immediately affects demand, as they can specify a lagged response, they do normally assume that the structure of this response follows a stable pattern over time and that it reflects a continuous and reversible functional relationship between prices and demand.

These assumptions can be questioned for reasons that have already been discussed. In particular, since the major impact of changes in material prices occurs indirectly through the development of new technology, neither the magnitude of the response to a change in price nor its lagged structure remains constant over time. On some occasions the response is negligible, while on others price changes provoke innovative activity that drastically alters consumption patterns. In addition, complete reliance on the use of material prices to capture the effects of substitution is certain to miss the

impact of those substitutions that are motivated by considerations of performance and quality or that occur at later stages of production where material costs are trivial. In tin-using industries such substitutions have often been important, and have at times substantially increased the use of tin despite its rising price.

Thus, only the last of the three groups of forecasting techniques appears to offer any hope of adequately accounting for the effects of substitution in predicting future mineral requirements, at least over the longer term. This is the only approach that can take full account of the abrupt and inconsistent, yet major, effects of material substitution that are caused not only by shifts in material prices, but also by changes in technology, government regulations, and other factors as well. Even here, in light of the inherent uncertainties involved in predicting the future course of new technology, government regulations, and the other factors affecting substitution, wide margins of error should be expected.

To some extent this pessimistic assessment can be modified when forecasting the requirements for a material in all of its end uses together, rather than requirements in specific end uses. Aggregation in this situation may lead to better results, because the large distortions in consumption caused by material substitution in individual end uses may to some extent cancel out. This is particularly likely to be the case for materials used in many applications. Still, given the pervasive influence of material substitution, the reliability of forecasting techniques that do not explicitly consider material substitution and take account of its discontinuous and abrupt nature must be questioned.

In summary, an examination of various uses of tin over the past several decades in the United States suggests that substitution, when defined broadly, is a major force shaping and altering material consumption patterns. Substitution, in turn, is driven primarily by technological change, and to a somewhat lesser extent by relative material prices and government actions. Material prices influence substitution largely by encouraging the development of new technologies that conserve or replace materials whose price is high and expand the uses of materials whose price is low.

These findings, to the extent that they reflect the nature of material substitution in general, have a number of implications. They call into question the common assumptions regarding the reversibility, continuity, and intertemporal stability of the demand curve, and suggest that the relationship between the demand for a material and its price is much more complex than is often assumed. In this regard, they raise the possibility that some of the methods now being used to estimate material demand functions and to analyze commodity markets may not be appropriate.

In addition, the findings imply that substitution greatly complicates the forecasting of future material requirements. If reliable long-term predictions are possible at all, they require that the future effects of material substitution be explicitly assessed. This cannot be done on the basis of past trends of historical relationships between material prices and demand. The findings also indicate that material substitution is a major constraint on the exercise of market power. Substitution can force firms in even highly concentrated indusstries to behave in a competitive manner. It also undermines the viability of cartels that the governments of mineral exporting countries might wish to create, and thus reduces the prospects for such collusive efforts. Finally, while cautioning that substitution may respond too slowly to changes in material prices to alleviate temporary shortages due to war, embargoes, cyclical demand surges, and strikes, the findings suggest that material substitution has a major contribution to make in the long-run struggle to prevent persistent shortages caused by the depletion of mineral resources.

PART II

CASE STUDIES

2

Beverage Containers

Frederick R. Demler

The first "soft drinks" were the mineral waters found in nature, and the first "beer" was brewed as long as 6,000 years ago from the grains raised by a society slowly giving up its nomadic existence. The containers for these drinks were crude—wooden kegs and earthenware vessels. Glass jars and bottles followed later of course, but it was not until 1809 that the tinplate metal can was invented by Nicholas Appert, who was seeking a way to preserve provisions for Napoleon's long war campaigns.

It was nearly another century before machine-made, mass-produced bottles were developed and 1935 before the first commercially acceptable tinplate beer container was introduced by American Can. A generation reared on throwaways might be startled to learn that commercial packaging of soft drinks in metal containers began only in 1953.

This chapter examines tin use in such beverage containers and the factors that have affected the amounts and distribution of its application. It looks at the apparent determinants of use and then examines the causes of changes in tin consumption from 1950 to 1977.

Apparent Determinants

The amount of tin used in beverage containers has fallen sharply over the past three decades despite a rally from 1958 to 1968. Tin consumption in beer and soft drink containers went from 6,300 tons in 1950 to 1,600 tons in 1977 (see figure 2-1), even though in 1977 soft drink cans used 70 percent of the tin that went into beverage containers.

Intensity of use has fluctuated also, as figure 2-2 shows. In 1950, 2.60 ounces of tin were used per barrel output of beer and none went into soft drink cans. This figure declined

over the next twenty-seven years while soft drink use rose to 0.53 ounces (in 1969) and then also declined.

The amount of tin used in any particular year t to produce beer or soft drink containers depends on five apparent determinants, as the following identity indicates:

$$Q_t = a_t\, b_t\, c_t\, d_t\, B_t$$

where: Q_t = tons of tin used for the beverage considered during year t.

a_t = proportion of the beverage that is packaged in bottles or cans during year t (as opposed to being shipped in bulk containers, such as kegs).

b_t = proportion of the packaged beverage that is put into tin cans (as opposed to aluminum cans, tin-free steel cans, and bottles) during year t.

c_t = number of average volume tin cans required per barrel of beverage in year t.

d_t = weight in tons of the tin contained in the average tin can used to ship the beverage in year t.

B_t = barrels of the beverage shipped in year t.

For example, 235.8 million barrels of soft drinks were produced in 1977, of which 78.9 percent were shipped in packaged containers. Tin cans, in turn, accounted for 14.48 percent of all packaged soft drinks. Each barrel of soft drink required 330.67 12-ounce cans, the average size of tin can used for soft drinks in 1977, and each of these cans contained 1.27×10^{-7} tons of tin. Altogether, 1,128 tons of tin were consumed in making tin cans for soft drinks in 1977, which

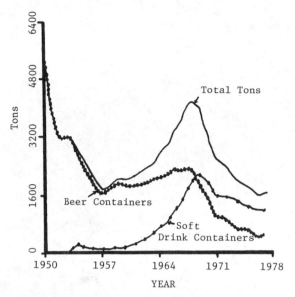

Figure 2–1.   Tin consumed in beer and soft drink containers, 1950–77. [From tables 2–1 and 2–2.]

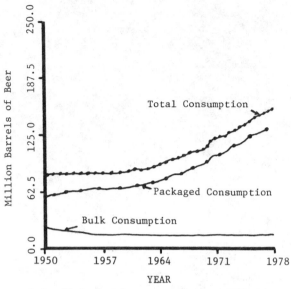

Figure 2–3.   Estimated beer consumption, 1950–77. [From table 2–1.]

as indicated by the above equation, is equal to: (235.8×10^6) (0.789) (0.1448) (330.67) (1.27×10^{-7}).

Trends in the apparent determinants of tin consumption in beer and soft drink containers over the 1950-77 period are given in tables 2-1 and 2-2. The rest of this section examines these trends.

barrels of soft drinks, whereas in 1950, 84 million barrels of beer were sold compared with only 48 million barrels of soft drinks. As shown in figures 2-3 and 2-4, beer consumption increased at an average annual rate of only 2.2 percent whereas soft drink consumption increased at an average annual rate of 5.8 percent.

Beverage Consumption

The amount of beer consumed by Americans appears to be declining in relation to soft drinks. In 1977, the United States consumed 155 million barrels of beer and 235 million

Packaged Containers' Market Share

The proportion of beer production shipped in packaged containers has increased over the postwar period, as shown in figure 2-3, while bulk shipments have remained about the

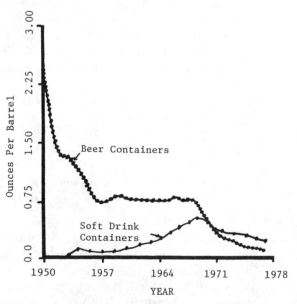

Figure 2–2.   Tin consumed per barrel output of beverage, 1950–77. [From tables 2–1 and 2–2.]

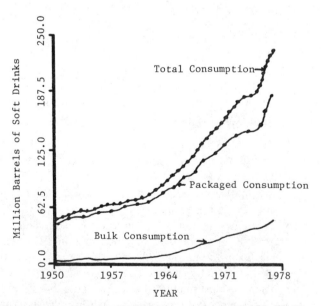

Figure 2–4.   Estimated soft drink consumption, 1950–77. [From table 2–2.]

Table 2-1. Apparent Determinants of Tin Consumption in Beer Containers, 1950–77

Year	Beer Consumption[a] (millions of barrels)	Proportion of Consumption Shipped in Packaged Containers (percent)	Proportion of Packaged Shipments in Tinplate Cans (percent)	Number of Average Size Tinplate Cans Per Barrel[b]	Tin Content of Average Size Tin-plate Cans[c] (10^{-7} tons)	Tin Consumption in Beer Containers[d] (tons)
1950	84.3	72	26	331	12.22	6278
1951	84.4	74	22	331	8.23	3744
1952	85.7	75	24	331	6.03	3099
1953	86.0	78	29	331	5.79	3163
1954	87.3	74	31	331	4.19	2786
1955	85.5	78	34	331	3.35	2487
1956	86.4	78	35	331	2.51	1971
1957	85.1	79	36	331	2.08	1683
1958	84.8	80	37	331	2.08	1745
1959	86.4	82	39	331	2.08	1916
1960	89.7	80	37	331	2.08	1835
1961	88.7	82	36	331	2.08	1817
1962	91.5	82	36	331	2.06	1852
1963	92.3	84	38	331	1.95	1900
1964	97.1	84	40	331	1.89	2021
1965	101.2	83	40	331	1.89	2063
1966	103.2	85	42	331	1.86	2272
1967	109.3	84	41	331	1.78	2194
1968	109.8	88	40	331	1.78	2300
1969	114.9	89	31	331	1.78	1902
1970	125.3	86	26	331	1.71	1584
1971	126.3	89	18	331	1.60	1057
1972	132.8	87	17	331	1.52	974
1973	135.8	90	12	331	1.46	716
1974	144.4	89	10	331	1.45	597
1975	148.7	89	9	331	1.44	576
1976	150.6	89	11	331	0.88	428
1977	155.5	91	12	331	0.86	494

Sources: Beer consumption and the proportion of beer consumption that is packaged are from: U.S. Brewers Association (various years); and Weinberg (1979). The proportion of packaged shipments in tinplate cans is estimated on the basis of information in Way (1979). The average size of the tinplate beer can over the entire period, as discussed in the text, is assumed to be 12 ounces based on information from the U.S. Bureau of Census (various years); *Modern Packaging* (various issues); personal interviews with industry spokesmen; the U.S. Brewers' Association (various years); Way (1979); and Weinberg (1979). The tin content of the average size tinplate beer can is estimated using data from *Modern Packaging* (various issues); *Modern Metals* (various issues); *Beverage World* (various issues); *Tin International* (various issues); the U.S. Brewers' Association (1979); and personal interviews with industry spokesmen.

[a]A barrel contains 31 gallons or 3,968 ounces.

[b]The average size tinplate beer can, for reasons discussed in the text, is assumed to remain at 12 ounces over the 1950–77 period.

[c]Net tin weight of the most metal efficient tinplate container available commercially.

[d]Figures shown for tin consumption underestimate actual consumption for two reasons. First, they include only the net weight of beverage containers, and so exclude the tin in the tinplate scrap generated in production. Second, they are based on the weight of the most metal efficient container in use. The underestimate is not believed to exceed 20 percent for any year. Since data on the weight of the average metal efficient can and on gross consumption are not available, actual tin consumption cannot be calculated.

same. In contrast to beer, the use of bulk containers for soft drinks has increased steadily, so that by 1977 over 20 percent of U.S. consumption was shipped this way.

Tinplate Can's Share of Packaged Containers

In 1950, beer and soft drinks were packaged in either glass bottles or three-piece tinplate cans. By 1977, the variety of containers had expanded considerably, and included glass bottles (both returnable and nonreturnable), three kinds of steel cans (three-piece tinplate, two-piece tinplate, and tin-free steel), aluminum cans, and plastic bottles. Beer and soft drinks shipped by container type are shown in figures 2-5 and 2-6, respectively, for the 1950–77 period.

BEER CONTAINER MIX. In 1950, the returnable glass beer bottle accounted for 70 percent of packaged beer, the three-piece tinplate for 26 percent, and the one-way bottle for 4 percent. By 1977 these market shares had changed radically. The glass bottle and the three-piece tinplate together accounted for only 42 percent of total packaged consumption, while the aluminum, tin-free steel, and two-piece tinplate cans had captured 41, 7, and 10 percent of the beer container market, respectively.

The returnable glass bottle, a staple container in the beer industry prior to 1950, declined in use from a high of 42.6 million barrels in 1952 to 17.2 million barrels in 1977, partly because of the rise of the one-way bottle, which began in

Table 2-2. Apparent Determinants of Tin Consumption in Soft Drink Containers, 1950–77

Year	Soft Drink Consumption[a] (millions of barrels)	Proportion of Consumption Shipped in Packaged Containers (percent)	Proportion of Packaged Shipments in Tinplate Cans (percent)	Number of Average Size Tinplate Cans Per Barrel[b]	Tin Content of Average Size Tinplate Cans[c] (10^{-7} tons)	Tin Consumption in Soft Drink Containers[d] (tons)
1950	48.5	94	0	331	—	0
1951	50.5	92	0	331	—	0
1952	54.8	92	0	331	—	0
1953	57.0	91	0[e]	331	4.89	38
1954	56.9	91	3	331	4.19	201
1955	61.2	91	2	331	3.35	112
1956	63.9	91	2	331	2.51	78
1957	65.9	91	2	331	2.08	77
1958	65.8	90	3	331	2.08	83
1959	71.8	90	4	331	2.08	112
1960	71.5	91	6	331	2.08	168
1961	73.8	89	8	331	2.08	253
1962	80.7	89	9	331	2.08	341
1963	87.1	90	11	331	2.08	425
1964	94.3	88	14	331	2.06	577
1965	101.8	87	17	331	2.05	770
1966	113.8	86	17	331	1.96	1104
1967	119.5	83	22	331	1.93	1386
1968	134.4	84	26	331	1.89	1840
1969	141.0	83	29	331	1.88	2133
1970	149.9	82	26	331	1.79	1912
1971	162.3	81	21	331	1.73	1555
1972	171.4	81	19	331	1.70	1508
1973	182.5	80	19	331	1.59	1458
1974	183.8	79	18	331	1.54	1349
1975	189.5	79	16	331	1.53	1222
1976	214.3	80	15	331	1.36	1160
1977	235.8	79	14	331	1.27	1128

Sources: Soft drink consumption, the proportion of soft drink consumption that is packaged, and the proportion of package shipments in tinplate cans are estimates based on information from: the National Soft Drink Association (various years); the U.S. Bureau of Census (various issues); *Modern Packaging* (various issues); and Way (1979); and personal interviews with industry spokesmen. The sources for the number of average size tinplate cans per barrel of soft drink and the tin content of the average size tinplate soft drink can are the same as those shown for beer in table 2-1.

[a]A barrel contains 31 gallons or 3,968 ounces.

[b]The average size tinplate can, for reasons discussed in the text, is assumed to remain at 12 ounces over the 1950–77 period.

[c]Net weight of the most metal efficient tinplate container available commercially.

[d]Figures shown for tin consumption underestimate actual consumption for two reasons. First, they include only the net weight of beverage containers, and so exclude the tin in the tinplate scrap generated in production. Second, they are based on the weight of the most metal-efficient container in use. The underestimate is not believed to exceed 20 percent for any year. Since data on the weight of the average metal efficient can and on gross consumption are not available, actual tin consumption cannot be calculated.

[e]The proportion of packaged shipments in tinplate cans for 1953 was 0.45 percent.

1960 and continued to grow rapidly, reaching a high of 39.0 million barrels in 1977. By 1977 one-way glass bottles outnumbered returnables by more than two to one.

The three-piece tinplate beer container grew rapidly from 1950 through 1967 when it was the most widely accepted beer container. After 1968, however, its use declined sharply.

The aluminum can was the principal container in 1977. Following its introduction in 1958, its share of the beer container market grew at a surprising annual rate of 33 percent, surpassing the three-piece tinplate can in 1972, and both the returnable and the one-way glass bottle in 1973.

Two fairly recent entrants into the beer can market include the tin-free steel can and the two-piece tinplate can. The tin-free steel can grew rapidly from its inception in 1967

and in the early 1970s was the most widely used beer container on the market. In the mid-1970s, however, its popularity declined as fast as it had risen. The other recent entrant, the two-piece tinplate can, entered the market only in 1971 and expanded in usage to 14.2 million barrels in 1977. It surpassed the three-piece tinplate can in 1975 and the tin-free steel can two years later.

Thus, over a twenty-seven year period, the package mix of beer containers in the United States changed substantially and rapidly. The returnable glass bottle, initially the principal beer container, was supplanted in the mid-1960s by the three-piece tinplate can, then the tin-free steel can, and finally the aluminum can, which by the end of the period was the preferred container. In addition, the nonreturnable

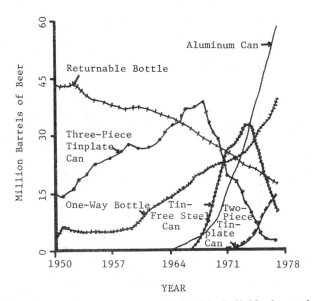

Figure 2–5. Barrels of beer packaged in individual containers, 1950–77. [From Way (1979), Glass Packaging Institute (1978), Weinberg (1979), U.S. Brewers Association (various issues), Demler (1980).]

bottle, although it has never dominated the market, has grown increasingly important.

SOFT DRINK CONTAINER MIX.   In 1950, soft drinks were packaged in returnable and nonreturnable bottles, as shown in figure 2-6. The former accounted for 44.4 million barrels, and the latter for 0.1 million barrels. In contrast to beer, the soft drink returnable bottle has continued to dominate

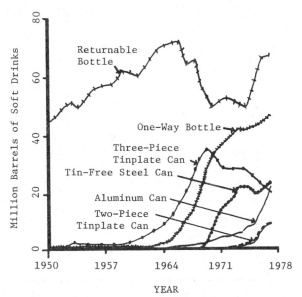

Figure 2–6. Barrels of soft drinks packaged in individual containers, 1950–77. [From Way (1979), Glass Packaging Institute (1978), National Soft Drink Association (various issues), Demler (1980).]

the container market, although with fluctuations, as the figure shows.

The nonreturnable glass bottle has continued to grow from 0.1 million barrels in 1950 to 50.8 million barrels in 1976. Between 1965 and 1977, its growth averaged 25 percent per year.

The three-piece tinplate can, on the other hand, has not been as successful. It went through a sluggish growth period when it was introduced (1954–59), accelerated for nine years, and finally declined in use (1970–77).

While the aluminum can is a recent entrant, it has grown rapidly, capturing a significant share of the soft drink market—20.8 million barrels in 1977.

Other recent entrants into the soft drink packaging business include tin-free steel cans and two-piece tinplate cans. The tin-free steel can, introduced in 1970, grew almost 23 percent a year until by 1977 it was the third most popular soft drink container. The two-piece tinplate can has so far had only a minor role in soft drink packaging, but its recent substantial growth suggests that it will become increasingly important in the future. A final container for soft drinks is the plastic bottle. (It is not shown in figure 2-6 because its use is so recent.) Although it was tested in 1970, the first big year of commercial use for this container was 1977 when an estimated 4 million barrels, or about 1.7 percent of total soft drink sales, were shipped in plastic bottles.

The soft drink packaging mix can be divided into two periods. From 1950 to the mid-1960s, there was little change in the mix of containers, for the returnable bottle dominated the market. In the second period, the mid-1960s to 1977, new containers entered the market and there was a good deal of fluctuation. As use of the nonreturnable bottle and the three-piece tinplate can grew, the returnable bottle declined and the introduction of the aluminum, tin-free steel, and more recently, the two-piece tinplate cans reduced use of the three-piece tinplate can. Finally, the returnable bottle, despite the competition from the new containers, showed signs of new growth during the 1970s.

DIFFERENCES AND SIMILARITIES.   While containers for beer and soft drinks use basically the same material inputs (except for plastics), the mix of containers over the period under study differs substantially. The returnable bottle, for instance, remains the principal soft drink container, while beer is primarily packaged in the nonreturnable bottle, the three-piece tinplate can, the aluminum can, and the tin-free steel can. The new containers were also introduced at different times:

	Beer	*Soft drinks*
Three-piece tinplate can	1935	1953
One-way bottle	1935	1948
Aluminum can	1958	1967

	Beer	*Soft Drinks*
Tin-free steel can	1967	1969
Two-piece tinplate can	1971	1973

A related difference concerns the rate at which new containers are accepted. The three-piece tinplate can accounted for 26 percent of the beer market in 1950 and 37 percent in 1963. Soft drinks were all packaged in glass containers in 1950, and the three-piece tinplate can had captured only 8 percent of this market by 1963. In 1977, the aluminum can was the principal beer container, with 41 percent of this market, but only the fourth most popular soft drink container, with 11 percent of that market.

The more rapid adoption of new containers in the beer market led to the earlier demise of other containers. The three-piece tinplate can began its descent in the beer market in 1967, but not until 1969 in the soft drink market. The tin-free steel can reached its peak in the beer market in the mid-1970s, but continued to expand in the soft drink industry through 1977.

Average Size of Tinplate Cans

The number of cans required to ship a barrel of beer or soft drink depends on the average volume of the cans used. For metal beer containers, which include tin-free steel and aluminum as well as tinplate cans, the average volume over time can be estimated by dividing the total volume of beer packaged in metal cans by the number of metal cans. As figure 2-7 indicates, the average volume remains fairly constant at 12 ounces over the period. Its mean value is 12.0 ounces and its standard deviation only 0.25 ounces. No trend, either upward or downward, is apparent.

The exact size distribution of metal soft drink containers is not known, though over 95 percent of soft drinks filled through the years are believed to have been in 12-ounce cans and the National Soft Drink Association (1978) assumes all cans are this size in determining can sales.

Thus, the average size metal beverage container is quite close to 12 ounces, and has remained at approximately this level over the last three decades. This suggests that changes in the average volume of tinplate cans have had little effect on tin consumption.

Tin Content of Tinplate Cans

The total amount of tin used in beer and soft drink cans may vary over time for two reasons. First, as just discussed, the size and average volume of beer and soft drink cans may change. Larger cans use less tinplate per unit volume of beverage than smaller cans. However, since the average volume of tinplate beverage cans has remained quite stable over time, this consideration has not been of much significance.

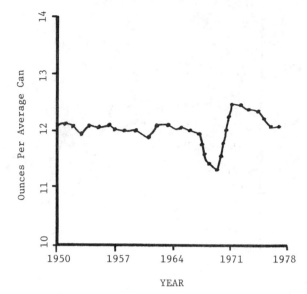

Figure 2–7. Average volume of beer cans. [From Weinberg (1979), Way (1979).]

Second, the amount of tin contained in a can of any particular size may change. Figure 2-8 shows that the tin content of beer and soft drink cans declined in general by 90 percent or more between 1950 and 1977. In 1950, the tin content of the 12-ounce three-piece tinplate can was about 2.77 pounds per thousand cans. By 1977, the most metal-efficient can (i.e., the can using the least amount of metal) was consuming only 0.16 pound of tin per thousand cans.

This sharp reduction in part reflects the substitution of aluminum and tin-free steel for tinplate in the ends of tinplate beverage cans. From 1950 to 1958, all metal beverage con-

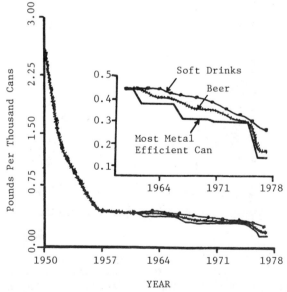

Figure 2–8. Tin content of tinplate beverage cans. [From tables 2–1 and 2–2.]

tainers had tinplate tops and bottoms. When aluminum beer cans were introduced in 1958, they also had aluminum top ends. In 1961, producers of the staple three-piece tinplate beer can started to fabricate tinplate cans with aluminum top ends (Demler, 1980), and by 1969 all producers had made the switch to aluminum ends. Thus, during the 1960s, there was a changeover from the tinplate "key punch" type ends to the "E-Z open" aluminum pull-tab type ends.

In the soft drink market, the switch started later and took somewhat longer. The first aluminum soft drink top ends were introduced in 1965 and an estimated one million were used by Coca Cola, Pepsi Cola, and 7-Up. The transition to the "E-Z open" aluminum end was completed by 1976.

When tin-free steel cans for beer and soft drinks came in (1967 and 1969, respectively), tin-free steel also replaced tinplate on the bottom ends of three-piece tinplate cans. It is not clear how the transition took place, but by 1973 for beer and 1976 for soft drinks, all three-piece tinplate cans apparently had a tin-free steel bottom (personal interview, American Can).

Comparative Effects of Apparent Determinants

The changes in the apparent determinants of tin consumption just discussed are responsible for the changes in the quantity of tin used for beer and soft drink containers in the United States. To assess the relative importance of the determinants that tended to increase tin usage, individual effects were estimated by calculating the increase in tin consumption each alone would have caused had all the other apparent determinants (including those tending to reduce tin usage) remained unchanged between 1950 and 1977.

The results are shown in figure 2-9. The first column for beer indicates the amount of tin—6,278 tons—used to make beer cans in 1950. The second column indicates the amount of tin—14,679 tons—that would have been consumed in 1977 had only those apparent determinants changed whose influence increased the use of tin between 1950 and 1977. The individual effects of these determinants are also shown in this column: The rise in beer consumption, had it alone changed, would have increased tin consumption by 5,299

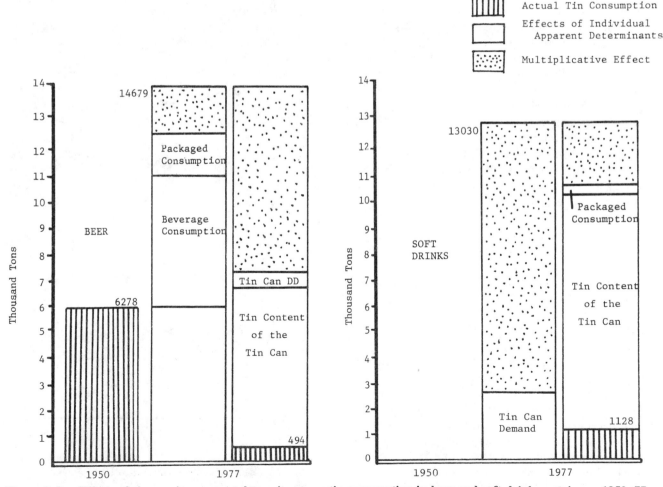

Figure 2–9. Effects of changes in apparent determinants on tin consumption in beer and soft drink containers, 1950–77.

tons, while the growth in the share of the total container market held by packaged containers, had it alone changed, would have expanded tin usage by 1,683 tons. When two or more apparent determinants tend to increase tin consumption, they produce a multiplicative effect that must be added to their individual effects. For example, if beer consumption increased by 50 percent and the share of packaged containers in the total beer container market grew by 50 percent, total tin consumption in beer containers would rise by 125 percent, not just 100 percent. The upper area of the second column in figure 2-9 indicates this multiplicative effect—1,420 tons—for the two apparent determinants stimulating tin consumption in beer containers.

The third column indicates the actual amount of tin—494 tons—consumed in beer cans in 1977, along with the relative influence of those apparent determinants reducing the use of tin. The individual effect of each of these negative determinants is estimated by calculating the amount by which tin consumption in 1977 would have increased had a particular apparent determinant remained unchanged from 1950, assuming all other determinants changed as they did.

In the case of beer containers, the negative apparent determinants include the tin content of tinplate cans and the decline in the tinplate can's share of the packaged container market. The reduced tin content of the tinplate can had an individual effect of about 6,537 tons and overshadowed the negative impact of the reduced tin can market share, the individual effect of which was a loss of 537 tons. The multiplicative effect, which shows the amount that tin consumption would have increased above the individual effects of these two negative apparent determinants had neither changed, is equal to about 7,111 tons, and again is represented by the shaded portion of the column. It is large primarily because of the substantial decline in the tin content of beer cans, so most of the reduction in tin use can be attributed to this determinant.

A similar portrayal for soft drink containers is shown in figure 2-9. Since tinplate cans were not used for soft drinks in 1950, the first column has a height of zero. The second column shows the rise in tin consumption—2,679 tons—due solely to the introduction of the tinplate can into the soft drink market and the resulting share of the packaged container market it captured. The increase in soft drink consumption also stimulated the use of tin. However, its effect is shown in the shaded area of column two as part of the multiplicative effect, since without the introduction of the tinplate can, it would not have affected tin usage. The combined effect—13,030 tons—indicates the amount of tin that the manufacture of soft drink containers would have required had only the positive determinants changed.

The third column shows the actual amount of tin—1,128 tons—used in soft drink containers during 1977, along with the negative effects of the reduction in the packaged containers' share of the total soft drink market (221 tons) and

the decline in tin content of tinplate cans (9,753 tons). Again, these are individual effects; the multiplicative effect is equal to 1,928 tons.

Material Prices and Technological Change

This section examines the major underlying factors responsible for the changes in the apparent determinants of tin consumption. Material prices are discussed first. However, because changes in material prices are found to have their greatest impact through the incentives and stimulus they provide for the development of new products and processes, this section also considers the effects of technological change on tin consumption in beer and soft drink containers.

According to economic theory, prices should have a major influence on material substitution and consumption. In this instance, the relevant prices include the price of tin as well as those of substitute and complementary materials. The complementary material is steel (in tinplate) and the substitute materials are chrome (tin-free steel), aluminum (aluminum cans), and glass (comprised of mainly silica sand and soda ash).

Indices of changes in these prices, measured in constant (1977) dollars over the 1950–77 period, are shown in figure 2-10, while changes in the price of tin relative to chrome, aluminum, glass, and steel prices are indicated in figure 2-11. Tin prices experienced the greatest increase over the entire period, with 1977 prices approximately 270 percent higher than those for 1950. Following tin were steel, aluminum, chrome, and finally glass (whose price actually declined in real terms). During the 1950s, tin prices rose

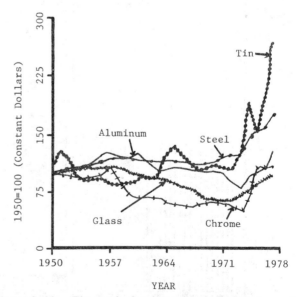

Figure 2–10.   Tin, steel, aluminum, chrome, and glass prices.
[From *American Metal Market* (1979).]

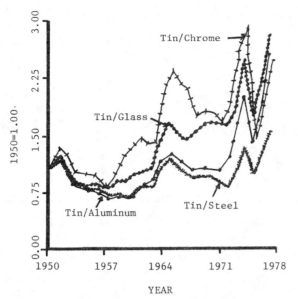

Figure 2–11. Price ratio of tin over steel, aluminum, chrome, and glass. [From *American Metal Market* (1979).]

more slowly than the other materials and increased rapidly only during the 1960s and 1970s.

As the rest of this section indicates, the effect of price changes on the apparent determinants of tin consumption is not always direct and consistent. In addition, their influence varies greatly, depending on the apparent determinant.

Beverage Consumption

The consumption of beer and soft drinks grew at average annual rates of 2.3 and 6.0 percent, respectively, over the 1950–77 period, while the real price of tin increased at a rate of 3.75 percent. Between 1950 and 1963, when the real price of tin actually declined, the growth rates of beer and soft drink consumption were also relatively low, which is not what one would expect if the price of tin were an important underlying factor affecting beverage consumption. (See table 2-3.)

This finding, however, is not that surprising when one considers the proportion of total costs accounted for by tin at various stages or steps in the production of a can of beer or soft drink. This information is provided in table 2-4 for 1978. At that time, the price of tin was 5.25 dollars per pound, or about 0.83 dollar per thousand cans. Steel costs,

at a price of 282.88 dollars per ton, ran about 9.07 dollars per thousand cans, so tin constituted about 8.4 percent of total raw material costs. At subsequent stages of production, the table shows that tin was responsible for a declining share of total costs—accounting finally for less than 1 percent of the retail price of soft drinks. Since beer is more expensive than soft drinks, the tin costs as a percentage of the total costs of filled beer cans are even lower.

The small contribution of tin to the total costs and the final beverage price paid by the consumer means that changes in the price of tin have a negligible effect on beverage consumption. The same holds true for the other raw materials used to manufacture beverage containers. Trends in population growth, the age distribution of the population, personal income, and consumer preferences are much more important in determining consumer behavior.

Packaged Containers' Market Share

The share of the total container market supplied by packaged containers increased for beer and decreased for soft drinks over the 1950–77 period. The increase for beer occurred at a time of relatively rapid increases in the price of tin, and thus suggests that the latter is not a major factor affecting the use of packaged containers. While the decrease for soft drinks is consistent with the hypothesis that the price of tin has some influence, later sections of this chapter indicate that the relative decline of packaged containers is largely the result of changes in consumer preferences and other factors completely unrelated to tin and other raw material prices.

Tinplate Can's Share of Packaged Containers

The mix of packaged containers can be divided into returnable and nonreturnable containers and into glass and metal containers. The latter can be further broken down into tinplate, tin-free steel, and aluminum cans.

The bottle is the only returnable container. The major factors affecting its use are: (1) overall cost savings to consumers compared with nonreturnable containers, (2) concern over the environment and conservation of materials, (3) the market structure of the beverage industry, (4) government deposit laws and regulations, and (5) the inconvenience and cost of returning, cleaning, and reshipping the bottle. These factors are all considered in later sections. The price of tin may have some effect on the overall cost savings associated with returnable bottles but the effect is small because tin is such a minor fraction of the final container cost.

The competition between glass and metal nonreturnables is partly influenced by costs, specifically those incurred by the packager. In addition to the purchase price of the empty containers, they include breakage losses, investment in filling equipment, filling efficiencies, labor, transportation,

Table 2-3. Average Annual Rates of Growth of Beer and Soft Drink Consumption and Tin Prices (percent)

	1950–1977	1950–1963	1964–1977
Beer Consumption	2.29	0.65	3.42
Soft Drink Consumption	6.03	4.28	5.11
Tin Prices (Constant Dollars)	3.75	−0.64	5.85

Source: Demler (1980).

Table 2-4. Tin Costs as a Percent of Total Costs at Different Stages of Production (1978 Costs are Expressed as $/1,000 Soft Drink Cans)

Industry	Raw Material	Fabrication of Steel Products	Can Manufacturer	Beverage Producer (Soft Drinks)	Ultimate Consumer
Product	**Tin, Steel**	**Tinplate**	**Metal Can (2 pc. Tinplate)**	**Canned Soft Drink**	**Final Purchased Beverage**
Tin Cost	$ .83	.83	.83	.83	.83
Total Cost	$9.91[a]	22.97[b]	36.57[c]	106.67[d]	183.33[e]
Tin Costs as Percent of Total Costs	8.4	3.6	2.3	0.8	0.5

Sources: American Metal Market (1979); Placey (1978); National Soft Drink Association (1978); personal estimates.
[a]Includes steel billet price.
[b]Price of tinplate containing 118 pounds of steel and 0.25 pound of tin per base box (31,360 square inches).
[c]Cost of fabricated can.
[d]Selling price of canned soft drink from beverage producer.
[e]Average retail price of packaged soft drinks.

storage, and other indirect costs. The metal can (whether tinplate, tin-free steel, or aluminum) has a higher purchase price, but it also offers lower breakage, higher filling efficiencies (because of its wide opening), and lower transportation and storage costs (because of its light weight and compactness).

Even if one ignores these costs to the packager and considers only the price of containers, tin prices constitute simply too small a portion—2.3 percent according to table 2-4—of the total costs of an empty can to affect greatly the mix of metal and glass containers. Much more important are consumer tastes, convenience, and other factors.

The price of tin becomes more important in the competition among tinplate, tin-free steel, and aluminum for the can market. This is in part because these three containers are quite similar in weight, durability, convenience, and other considerations that strongly influence the packager's choice between glass and metal. Indeed, the average consumer is usually not even aware of the material composition of the beverage cans he uses.

Over the 1950–77 period, technological developments repeatedly affected the costs for empty cans—and in turn the battle for the metal can market. In the early 1950s, tinplate cans fabricated by the three-piece technology with a soldered side seam monopolized the market. Aluminum could not be economically soldered and thus was precluded from conventional canmaking technology. In 1958, Coor's Brewing and Beatrice Foods introduced the impact extruded two-piece aluminum can. While this technology allowed aluminum to be used as a canmaking material, three-piece tinplate cans retained their comparative advantage because of lower costs. In 1963, Reynolds Aluminum first produced an aluminum can using the draw-and-iron process, which proved to be the efficient two-piece technology. Although the price of aluminum increased between 1950 and 1963 while the price of steel remained fairly constant and the

price of tin fell, the two-piece canmaking technology enabled aluminum to compete effectively with tinplate. Between 1963 and 1977, the market share of aluminum consistently increased, even though over at least part of this period the price of aluminum relative to steel (the major raw material cost for tinplate) increased dramatically. As a result, by 1977 aluminum held 68 percent of the beer can market and 30 percent of the soft drink can market.

The aluminum challenge forced steel producers and canmakers either to improve upon the three-piece tinplate can or find an alternative container if they wished to maintain a share of the beverage can market. Steelmakers responded by introducing the tin-free steel can, a chrome-plated steel can with about 0.011 pound of chrome per thousand units. This compared to about 0.462 pound of tin per thousand three-piece tinplate cans. In dollar terms, the tin cost of tinplate cans was about 71 cents per thousand in 1967 and the chrome cost in tin-free steel cans about 1.4 cents. Tin-free steel, however, could not be economically soldered. Canmakers later introduced a welded tin-free steel can (Continental Can's Conoweld) and a cemented tin-free steel can (American Can's Miraseam). The use of these cans, which were less expensive than either the three-piece tinplate can or the two-piece aluminum can, rapidly expanded, as figures 2-5 and 2-6 indicated earlier. Canmakers essentially substituted the tin-free steel technology for the three-piece tinplate technology.

Aluminum canmakers were quick to realize the potential of tin-free steel and the threat it posed for them. In 1969, two years after tin-free steel cans first appeared, they introduced two-piece aluminum cans from the H19 alloy, a full hard temper material that was stronger and lighter than previous alloys. It could be fabricated into metal cans at less expense than previous aluminum containers and once again allowed aluminum to compete effectively for the beverage can market.

In response to this new challenge from the aluminum industry, steel producers and canmakers began experimenting with two-piece steel cans. Pure steel (blackplate), tin-free steel, tinplate, and other materials were tested for the two-piece technology. Blackplate, the least expensive material, lacked the desirable properties of corrosion resistance and lubricity. Lubricity refers to tin's ability to lubricate the exterior of a can and thus aid in the forming process. It is essential for proper forming of the can body in the two-piece process. Tin-free steel was also relatively inexpensive but again lacked lubricity. Only tinplate, owing to its tin content, had the necessary lubricity as well as a corrosion resistance, and so it became the material used to produce two-piece steel cans. These cans first appeared for beer in 1971 and for soft drinks in 1973.

Steel's late entry into two-piece technology was in part due to the higher capital and investment costs involved in fabricating two-piece steel cans. The production costs and capital costs per thousand units are shown in table 2-5 for aluminum and steel two-piece beer and soft drink cans. The coating, tooling, and labor costs, overhead, and depreciation are all higher for steel than for aluminum. Tinplate, on the other hand, is less expensive than aluminum sheet, so that net metal costs for tinplate cans are lower—22.45 dollars per thousand in 1978 compared to 27.87 dollars for aluminum.

Overall, according to table 2-5, the cost of producing an aluminum two-piece beverage container is about 12 percent greater than that for a two-piece tinplate container. Production costs, however, are not the only costs that enter into the beverage packager's decision. Aluminum cans weigh about 30 pounds and tinplate cans about 70 pounds per thousand, which means that aluminum cans have lower transportation costs. In addition, aluminum cans do not rust

or distort the taste of beverages, as tinplate cans are alleged to do, and they are easier and cheaper to recycle. Tinplate, however, can be processed using magnetic equipment. Thus, the decision to use tinplate or aluminum entails more than a simple comparison of production or material costs.

Since the introduction of two-piece tinplate cans for beer and soft drinks, no new three-piece beverage canmaking capacity has been built. The number of two-piece tinplate beer cans increased from 0.1 to 5.8 billion between 1971 and 1978, while the number of two-piece soft drink cans rose from 0.1 to 4.1 billion.

This discussion indicates that technological change has been an important ally in the battle for the beverage can market, favoring first one can and then another. Tin, steel, chrome, and aluminum prices have also influenced choices, though indirectly, by encouraging the development of particular innovations. The rise in the price of tin relative to that of chrome over the 1957–65 period, for example, stimulated interest in the new technology that produced the tin-free steel can in 1967. Similarly, the subsequent decline in the relative price of tin between 1965 and 1971 presumably encouraged the development of the two-piece tinplate can.

Material prices, however, so far have not directly affected the composition of the can market in the sense that a rise in the price of tin caused tinplate canmakers to switch to tin-free steel cans or to aluminum cans. Such substitutions have been precluded by the nature of the production technologies, which required expensive changes in procedures and equipment to switch from one material to another.

Tinplate was used in the three-piece soldered can, tin-free steel in the three-piece cemented or welded can, and aluminum in the two-piece can. The differences in production technologies, however, declined considerably for aluminum and tinplate canmakers with the introduction of the two-piece tinplate can. As a result, some canmakers are now investing in canmaking machinery (at an initial capital cost disadvantage of about 500,000 dollars) which can use either tinplate or aluminum. The first such line was constructed in 1976, and the downtime to switch material inputs is less than one day. So, for the first time in the history of the beverage can industry, material substitution in response to price changes can now take place quickly at existing facilities with little changeover costs. Historically, material prices could influence the type of new canmaking facilities built, and thus to some extent the mix of metal beverage cans. However, over much of the period examined, existing technology at the time new capacity was constructed tended to dominate the decision regarding the type of can to be produced. The prices of tin and other metals were of lesser importance.

Average Size of Tinplate Cans

As indicated in the preceding section, the average size of metal beverage containers has remained fairly constant over

Table 2-5. Costs of Producing Two-Piece, 12-Ounce Cans from Aluminum and Tinplate in 1978 (dollars per 1,000 cans)

Cost Element	Aluminum[a]		Tinplate[b]	
	Beer	Soft Drinks	Beer	Soft Drinks
Metal				
Starting	31.75	31.75	22.98	22.9
Scrap Credit	3.88	3.88	.53	.53
Net	27.87	27.87	22.45	22.45
Cleaning & Treatment	.30	.30	.10	.10
Interior Coating	.65	.90	.975	1.19
Tooling Cost	.15	.15	.35	.35
Labor Cost	4.93	4.93	5.11	5.11
Overhead Cost	2.33	2.33	2.41	2.41
Depreciation	3.58	3.58	4.20	4.20
Total Cost	39.81	40.06	35.60	35.81

Sources: Placey (1978); Richardson (1978).
[a]Made from aluminum sheet weighing 33 pounds per base box.
[b]Made from tinplate containing 118 pounds of steel and 0.25 pound of tin per base box.

the 1950–77 period. If tin prices had an important influence on the size of the containers, the latter should have decreased during the 1950s and then expanded during the 1960s and 1970s with changes in the price of tin. But this did not happen because consumer preferences, government regulations, socioeconomic variables, institutional considerations, and other factors are far more important than material prices in determining average can size.

Tin Content of Tinplate Cans

Before 1946, all beer cans were produced from hot-dipped tinplate. In the early 1950s the lower cost (*Tin*, 1952, p. 13) and more uniform coating of electrolytic tinning encouraged a switch to this process and by 1955 it completely replaced hot-dipped tinplate, resulting in a marked decline in the amount of tin used. Electrolytic tinning was commercialized during World War II to conserve tin supplies and after its inception there was extensive research on tin weight, corrosion protection, and expected shelf-life of products. Much of this research focused on the possibilities of differentially coated tinplate, in which the outside of the can received lighter coatings of tin.

In the 1930s, most metal cans required a "standard coke" of 1.5 pounds of tin per base box (a base box is a unit of area used in the tinplate industry and is equivalent to 31,360 square inches) or about 2.77 pounds of tin per 1,000 cans. With emphasis on cost reductions, can companies and steel producers experimented together to reduce tinplate raw material costs by matching product requirements with minimum tinplate quality (McKie, 1959, p. 267). By 1957, the standard steel beer can was manufactured with 0.25 pound tinplate (0.462 pound of tin per 1,000 cans), with equal tin coatings on the inside and outside. Enamel supplemented these minimum tin coatings to provide corrosion resistance and taste protection.

In 1961, the tinplate top end of the three-piece tinplate beverage container was supplanted by the aluminum flip top lid, further reducing tin content. In 1967, tin-free steel cans first appeared, and the lower cost tin-free steel was also used for the bottom end of the three-piece tinplate can. This caused tin weight per thousand tinplate cans to drop to 0.330 pound.

The figure remained constant until 1971 when two-piece cans were first fabricated from steel. By 1976, 0.25 pound tinplate was being used, and tin weight per thousand cans dropped to 0.159 pound. This last decline was the result of extensive research on the optimal relationship between enamels, tin coating, and product corrosion specifications, which was similar to research conducted during the 1950s.

This review indicates that the sharp decline in the tin content of tinplate cans over the last three decades was in large part the result of numerous technological developments. This raises the question of how much these inno-

vations were encouraged or stimulated by the relatively high price of tin, as well as the increase in the price of tin over this period. A precise answer to this question, of course, is not possible. And to some extent, the evidence is conflicting. For example, the greatest reduction in tin content occurred during the 1950s, when the price of tin was falling relative to the other materials used in beverage containers. Still, there is no question that steel producers and canmakers are strongly motivated in their research and development efforts by the desire to cut costs. The relatively high price of tin, plus the fact that it accounts for some 8 percent of the raw material costs of producing tinplate, makes research on ways to reduce tin requirements attractive. Among such efforts are electrolytic plating, lighter coated tinplate, enamel coatings, the tin-free steel bottom, and the two-piece tinplate can, all of which have helped reduce the tin content of the tinplate can.

It is likely that the high price of tin and increases in the price over time have affected the tin content of the tinplate can indirectly by increasing the returns to tin-saving innovations. There is little evidence to suggest, however, that changes in the relative price of tin have an immediate effect in causing canmakers to reduce the amount of tin or increase the amount of other materials. This is not surprising since at any particular time tin requirements are specified by existing technology and the particular canmaking equipment in use. The one exception is the dual canmaking facility that has recently been introduced and still accounts for only a small fraction of total canmaking capacity.

Nonprice Factors

The preceding section concluded that technological developments have greatly affected two of the more important apparent determinants of tin use—the share of the beverage container market held by the tinplate can and the can's tin content. Material prices, through their indirect effect on the level and direction of research and development, were also found to influence these two determinants. This section considers the major nonprice factors affecting the apparent determinants of tin consumption—demography and per capita income, consumer preferences, government regulations, and market structure and conduct in the container industry.

Demography and Per Capita Income

Between 1950 and 1977, consumption of beer increased by 84 percent and soft drink consumption rose 314 percent. The major factors responsible for this growth include a 43 percent expansion in the U.S. population, along with changes in the age and regional distribution of the population, and an increase in per capita income.

As table 2-6 indicates, per capita consumption of beer and soft drinks grew along with per capita income. With

Table 2-6. Per Capita Beverage Consumption and Income, Selected Years, 1950–77

Year	Beer Consumption (Gallons Per Capita)	Soft Drink Consumption (Gallons Per Capita)	Per Capita Income (Constant 1972 Dollars)
1950	16.8	9.9	2386
1955	16.0	11.4	2577
1960	15.0	12.1	2697
1965	15.9	16.3	3152
1970	18.6	22.8	3619
1975	21.5	26.7	4025
1977	22.6	33.7	4285

Sources: Brewers Association (various issues), National Soft Drink Association (various issues), U.S. Council of Economics (1980).

rising incomes, consumers have more funds to spend on discretionary items such as beer and soft drinks.

The age distribution of the consuming public also influences the demand for beverages. The type of beverage consumed generally follows a fairly consistent pattern over the life cycle of individuals. At an early age, milk, fruit juices, and water are consumed almost exclusively, and are then augmented by soft drinks in the preteen and teenage years. The young adult begins to substitute hot beverages (tea and coffee) and beer for milk, fruit juices, and to a lesser extent, soft drinks. In later years, distilled spirits and wine may be substituted for beer and soft drinks (Woodroof and Phillips, 1974, p. 21).

Table 2-7 shows the distribution of the population by three age groups—19 and under, 20 to 44, and 45 and older—for the years 1950, 1960, 1970, and 1977, as well as beer and soft drink consumption during these years. It suggests that the post-World War II baby boom was responsible for the high growth in soft drink consumption in the 1950s while beer consumption remained relatively flat. After 1960 and into 1970, beer consumption increased, partly as a result of the maturing of the baby-boom population. Soft drink consumption continued to increase, which reflects, to a degree, continuing consumption by the older segments of the population.

Soft drink producers during the 1970s also modified their marketing campaigns to encourage a growing proportion of older people to consume more of their product. Advertising campaigns, such as "Coke Adds Life" and "The Pepsi Generation," are not only aimed at the under-19 age group,

but also the over-20 age group. "The Pepsi Generation" is not a particular age group but rather portrays a psychological group to which anyone can belong. "Coke Adds Life" to everyone (especially those over 20 years of age). Thus, this modification in marketing strategy has somewhat altered the traditional concentration of soft drink consumption among the preteen and teenage group.

Another factor affecting beverage consumption is the geographic distribution of the population (Demler, 1980). The South tends to consume less beer and the West fewer soft drinks per capita than the rest of the economy. Among the reasons offered for the low per capita beer consumption in the South are the conservative and moralistic nature of the region, higher beer taxes, and the larger rural population. The rural environment, it is argued, has less tension and fewer opportunities for socializing.

The low per capita consumption of soft drinks in the West can probably be attributed to the more dispersed population and the problems this creates for the economic distribution of soft drinks.

In an effort to assess these four factors underlying changes in beverage consumption over the 1950–77 period, the following equation was estimated by regression analysis:

$$BevCsp_t = a_0 + a_1 TotPop_t + a_2 AgeDist_t + a_3 RegPop_t + a_4 PerCap_t + e_t$$

The dependent variable $BevCsp_t$ is the consumption of beer or soft drinks in year t in million barrels. The variable $TotPop_t$ is the total population of the United States in year t measured in millions of people. The variable $AgeDist_t$ is the percent of the population over 20 years of age in year t for beer, and the percent of the population between five and 44 years of age for soft drinks. $RegPop_t$ reflects the percent of the population in the South for beer in year t, and the percent of the population in the West for soft drinks in year t. $PerCap_t$ is the average per capita income in thousands of constant (1972) dollars in the United States in year t. The variable, e_t, is the disturbance term, and a_0, a_1, a_2, a_3, and a_4 are the parameters to be estimated.

parameter	sign	comments
a_1	positive	Increases in population should result in increases in beverage consumption.

Table 2-7. Beverage Consumption and Age Distribution of Population, Selected Years, 1950–77

Year	Beer Consumption (Million Barrels)	Soft Drink Consumption (Million Barrels)	Population (Millions of Persons)	Age Distribution of Population (percent of population)		
				19 and Under	20 to 44	45 and Above
1950	84.3	48.5	152.1	33.9	37.7	28.4
1960	89.7	71.5	180.6	38.5	32.2	29.3
1970	125.3	149.9	205.0	37.7	32.0	30.3
1977	155.5	235.8	217.0	33.6	35.4	31.0

Sources: Brewers Association (various issues), National Soft Drink Association (various issues), U.S. Council of Economic Advisers.

parameter	sign	comments
a_2	positive	As the percent of the population over 20 increases, the consumption of beer should increase; and as the percent of the population between five and 44 increases, the consumption of soft drinks should increase.
a_3	negative	As the percent of the population in the South and West increases, the consumption of beer and soft drinks, respectively, is expected to decrease.
a_4	positive	As per capita income increases, beverage consumption should increase.

The results of the regressions are shown in table 2-8. In all cases, the parameters have the expected signs and are generally significant. In the case of total population, statistical results indicate that a one million increase in population results in about a 0.94 million barrel increase in beer consumption and a 0.73 million barrel increase in soft drink consumption. A thousand-dollar increase in per capita income results in about a 7 million barrel increase in beer consumption and about a 43 million barrel increase in soft drink consumption. The significance of the coefficient for the AgeDist variable indicates that the percentage of the population that is 20 or older affects beer consumption, and that the percentage of the population between 5 and 44 affects soft drink consumption. The coefficients of the regional characteristics of the population, though not significant, do have the anticipated signs, suggesting that beer consumption is lower in the South and soft drink consumption in the West compared to the rest of the country. Finally,

the coefficients of determination for the two equations exceed 99 percent, which implies that the four independent variables can account for or explain almost all of the year-to-year changes in beer and soft drink consumption in the United States.

Consumer Preferences

Consumer preferences shaped by a host of physiological, sociological, and socioeconomic considerations comprise another set of underlying factors affecting beer and soft drink consumption, the mix of bulk and packaged containers, the share of the tinplate can in the packaged container market, and even the tin content of tinplate cans. Physiological considerations are primarily taste preferences, but also include dietetic, sweetness, and alcoholic preferences, along with thirst satisfaction. The sociological aspects center around the family, drinking habits, and social conformity. The socioeconomic effects relate to availability and convenience.

The importance of availability and convenience is stressed by many researchers. Weinberg (1971), for example, argues that the major factor affecting shifts in beer consumption is convenience packaging (one-way containers). "Cans and non-returnable bottles in six packs and easy-opening containers created new customers by increasing the number of appropriate occasions for drinking malt beverages and encouraging a high consumption per occasion." Weinberg goes on to say that "because of the handiness, compactness, easy cooling characteristics, and greater availability of the convenience package, the American consumer could enjoy malt beverages in more places and greater amounts."

Another aspect of convenience is the aluminum "E-Z open" flip top introduced in 1962. Prior to its commercialization, the sale of beer in metal cans was decreasing, from 9.1 billion cans in 1959 to 8.9 billion cans in 1962. Breweries who first converted to the aluminum top can saw their sales jump by 100 to 200 percent. This encouraged others to follow. As a consequence, all beer cans had aluminum flip top ends by 1969, and all soft drink cans by 1976. This switch completely eliminated tin from the top

Table 2-8. Regression Results Estimating Changes in Beverage Consumption in Response to Changes in Total Population, Age Distribution, Regional Characteristics of Population, and Per Capita Income Over the 1950–77 Period

Dependent Variable (In Million Barrels)	Coefficients of Independent Variables								
	Constant	TotPop	AgeDist	RegPop	PerCap	R^2	p	D.W.	n
Beer Consumption	−394.72* (7.95)	.943* (5.44)	580.4* (6.56)	−210.7 (.94)	7.1 (1.25)	.995	—	1.91	28
Soft Drink Consumption	14.76 (.24)	.728* (1.77)	464.6 (1.47)	−759.77 (1.35)	42.8* (4.23)	.996	.14	2.15	27

Notes: An asterisk indicates that the coefficient is statistically significant at the 95 percent probability level using a one-tailed test.

The dependent and independent variables are defined in the text. The t-statistics for the coefficients are shown beneath the coefficients in parentheses. Also shown are the coefficients of determination (R^2) for the equations, the correlation coefficient (p) between the first differences of the disturbance term estimated by the Hildreth-Lu procedure and used to correct for autocorrelation, the Durbin-Watson statistic (DW), and the number of observations. The Hildreth-Lu procedure was not used for the beer consumption equation since this equation does not appear to suffer from autocorrelation.

end of the beverage can, which in 1950 accounted for about 14 percent of the total tin content of tinplate cans.

A final convenience factor is the diffusion of vending machines. A study conducted for the American Medical Association using selected armed forces personnel showed that a positive relationship exists between the availability of soft drinks in vending machines and their total consumption. Soft drink retailers have been quick to make use of this marketing mechanism by substantially increasing the number of vending machines (Corplan Associates, 1966).

Another trend in the vending segment of the soft drink market is also of particular interest. Table 2-9 shows the number of vending machines operated in 1965 and 1977 by type of container. While the number of soft drink machines dispensing cans increased by more than 440 percent over this period, the number of bottle and cup machines actually declined. The obvious reason for this shift is that the metal containers do not break.

The physiological factor of taste preferences has created a highly inelastic market for glass containers among certain consumers. Glass is chemically inert and does not react with beer or carbonated soft drinks as metal containers do. Steel cans dissipate metal into the beverage at a higher rate than aluminum. As this may distort the taste, some consumers prefer their beer or soft drinks packaged in glass instead of metal and prefer aluminum over steel. The minimal taste-distorting properties of aluminum compared with steel have also helped aluminum increase its market share. Beverage producers first discovered the decline in metal migration when aluminum top ends were used on steel cans. Later, they realized that a can made entirely of aluminum minimizes taste distortion to an even greater extent.

The sociological factors affecting beverage consumption have had a significant impact on the proportion of both beer and soft drinks shipped in bulk and packaged containers. As noted earlier, the percent of beer consumed in bulk form has declined since 1950, while the opposite has occurred for soft drinks. Most of the beer shipped in bulk containers is consumed in taverns, along with a major portion of the returnable beer bottles. Since 1950, the trend toward home consumption has led to the demise of the local tavern, and a fall in bulk consumption and returnable bottle usage.

Growth in bulk soft drink consumption is due primarily to advances in mixing and filling equipment and the growth

of "fast-food" chains. The advances in soft drink premixed syrups and in fountain measuring equipment enabled retailers to produce a higher quality, more uniform soft drink, which enhanced the acceptability of bulk soft drinks. The rise of McDonald's, Burger Kings, Pizza Huts, and other fast food chains then created a rising demand for a bulk soft drink distribution system that was efficient and had minimum waste.

Postwar changes in living styles have encouraged the consumption of beer and soft drinks in packaged containers. The rise of backyard barbecues, picnics and other outings, family room activities, and television viewing have all stimulated the demand for packaged containers. These changes within the social setting of the home have also created a more informal atmosphere where beer and soft drinks have become more generally accepted, which has increased their overall consumption.

Consumption as a family unit has also encouraged the use of larger sized containers, which in turn has favored glass and plastic bottles. The larger sized soft drink bottles (48-64 ounces) increased in use by over 250 percent between 1974 and 1977, while an offsetting loss of 44 percent took place in 6-9-ounce containers. Larger containers use less material per ounce of beverage and so reduce material use. However, some researchers (R. S. Weinberg) have found that a trend toward larger containers may not actually result in a decline in material use. Evidence indicates that many individuals consume a six-pack of 16-ounce soft drink bottles as fast as a six-pack of 12-ounce soft drink bottles. This finding has encouraged soft drink bottlers to market their product in larger-sized containers, since larger containers apparently increase overall beverage consumption.

A final consumer preference factor relates to container identification. During the early part of the period examined, Coca Cola was primarily packaged in a patented 6-ounce green returnable bottle, while Miller "High Life" and other higher quality beers were marketed only in glass bottles. More recently, certain mineral waters, such as Perrier, are sold only in glass bottles. Containers present a certain image or quality which is used as a marketing tool. When beer and especially soft drinks were first marketed in tinplate cans, the beverage container industry promoted the image of being "modern" with metal cans. Finally, the use of the returnable bottle was to some extent encouraged by the belief among many beverage producers that it tends to lock the consumer into purchasing and repurchasing the same beverage. One-way containers, it is felt, give the consumer an added element of choice that could hurt sales.

Government Regulations

Over the period studied, the U.S. government has regulated segments of the beverage container industry and in the process altered the use of materials. Regulations were imposed

Table 2-9. Number of Soft Drink Vending Machines by Type of Container, 1965 and 1977

| Year | Type of Container (thousands) | | | |
	Can	Bottle	Cup	Total
1965	29.1	145.8	19.7	194.6
1977	128.6	56.7	9.6	194.9

Source: National Soft Drink Association (1977).

for strategic reasons during the Korean War and were similar to those of World War II. Restrictive packaging legislation has arisen in recent years, and there have been health and safety regulations throughout the period.

STRATEGIC REGULATIONS.   In 1950, the United States produced some 100 tons of tin while consuming over 115,000 tons from countries such as Indonesia, Malaysia, and Thailand. Owing to the distant location of the major producing regions, tin supplies were curtailed during World War II and the Korean War. Government policy during World War II was simply to restrict the use of metal containers in the domestic market. Breweries were considered nonessential, and thus were not allowed to purchase new canmaking equipment. During the Korean War, the use of metal containers for beverages was again restricted, and specific attempts were undertaken to conserve tin. The government coerced the beverage container industry to accept thinner container coatings despite higher costs in order to save tin (*Tin*, 1954, p. 4). These measures, however, later resulted in improvements in coating technology which reduced the costs and the tin content of the tinplate can.

The Defense Production Administration provided rapid tax amortization on capital invested in new tin-saving techniques during this early period. This substantially accelerated the changeover from hot-dipped tinplate to electrolytic tinplate. Government programs also stimulated research on dual-coated tinplate that eventually led to the use of 0.25 pound tinplate with extensive enamel coatings.

With the end of the Korean War, tin shortages ceased and the ban on the use of one-way containers in the domestic market was lifted. However, the search for techniques to use less tin that was initially sponsored by the government was continued by the private sector. American Can Company (1953) in unveiling its "Operation Survival" stated: "Although a truce marks the halt of war in Korea, the American Can Company is spearheading and winning another battle closely identified with the Red menace. It is a battle of research to free itself from dependence upon foreign tin supplies."

The program was aimed at Southeast Asian suppliers of tin where some 75 percent of productive capacity was located. American Can, in cooperation with the steel and container industries, studied the possibility of a "tin-less" can, and the economic feasibility of new materials and techniques. The research conducted in "Operation Survival" involved such efforts as using less tin in containers, eliminating tin in specific containers by substituting enamels, and replacing tin-lead solders with plastic cements. The final research program led in 1966 to the Miraseam chrome-coated beverage container, which eliminated tin from both the steel sheet and the bonding material.

World War II and the Korean War also stimulated the acceptance of one-way containers. The returning veteran was conditioned to using one-way containers while on active duty and continued to purchase beverages in these containers once back in civilian life. His role is considered to be one of the important factors underlying the growth of one-way containers in a society that historically was oriented toward the returnable bottle.

RESTRICTIVE PACKAGING LEGISLATION.   Proponents of restrictive packaging legislation have long argued that a return to reusable glass containers and the elimination of metal cans and one-way bottles would produce many benefits—less litter, lower solid waste collection and disposal costs, energy conservation, lower beverage prices, more employment, and reduced raw material consumption. With the move to "Keep America Beautiful," the issue of restrictive packaging legislation intensified, and in 1971, some 250 bills were pending in 38 state legislatures and 23 bills in Congress to restrict the use of one-way containers (Demler, 1980). Proposals ranged from the outright banning of one-way containers, to requiring specific deposits, taxes, subsidies, educational campaigns, and greater enforcement of litter laws (Environmental Protection Agency, 1972). Since that time, Oregon, Vermont, Maine, Michigan, Iowa, New York, Delaware, Massachusetts, and Connecticut have passed laws restricting the use of one-way containers. Other states and localities have imposed statutes affecting material use. In 1974, Senator Mark Hatfield introduced the "Nonreturnable Beverage Container Prohibition Act," to discourage the use of nonreturnable beverage containers in interstate commerce by suggesting that a five-cent refundable deposit be imposed upon all beer and soft drink containers.

The results of the Oregon and Vermont bills have been assessed by the proponents of packaging legislation, the beverage packaging and producing industries, and a number of outside observers. The findings cannot be generalized and applied to other states, since residents of Vermont and Oregon favored the returnable bottle system even before the bills were enacted. Michigan, on the other hand, has a packaging mix similar to the general economy, but it is still too early to analyze this case. The same is also true for Maine.

The beer package mix (cans, one-ways, returnables) and bulk consumption for these four states, before and after restrictive legislation, are shown in table 2-10. In general, the market share of cans tends to decline significantly after the deposit law is enacted. In Oregon, for example, cans accounted for 33 percent of the package mix in 1971, but only 5 percent by 1976. The one-way bottle also tends to decline, while the returnable bottle enjoys a significant increase in market share. There is also some evidence indicating that deposit laws tend to discourage the use of steel cans more than aluminum cans (Katz, 1978).

Packaging legislation also affects the demand for beverages and thus the demand for beverage containers. The

Table 2-10. Market Shares of Cans, One-Way Bottles, Returnable Bottles, and Bulk Containers in the Beer Industry Before and After Restrictive Packaging Legislation (percent of container market)

A. OREGON—Deposit Law Effective October 1972

	1971	1976	1977	1978
Cans	33	5	6	10
One-Ways	19	8	18	23
Returnables	25	63	53	45
Draught	23	24	23	22
	100	100	100	100

B. VERMONT—Deposit Law Effective September 1973

	1971	1976	1977	1978
Cans	39	31	28	27
Onc-Ways	45	35	41	44
Returnables	10	19	16	15
Draught	6	15	15	14
	100	100	100	100

C. MAINE—Deposit Law Effective January 1978

	1977	1978
Cans	38	32
One-Ways	50	52
Returnables	4	6
Draught	8	10
	100	100

D. MICHIGAN—Deposit Law Effective December 1978

	1977	1978	1979
Cans	64	56	33
One-Ways	14	18	12
Returnables	11	15	41
Draught	11	11	14
	100	100	100

Source: Katz (1979).

consumer loses convenience because of the return feature. In addition, fewer types of packages and brands of beverages are likely to be available in any given region, since national brewers cannot afford to ship and re-ship heavy returnable bottles to distant markets. In addition, the price of packaged beverages may increase because of the costs associated with transporting, storing, and refilling returnable bottles. For these and other reasons, sales often fall for both beer and soft drinks.

Another type of beverage container legislation bans detachable tops. Minnesota, Virginia, Maine, Michigan, Oregon, and Vermont had passed such laws by 1977 (*Tin International*, 1977, p. 3). To meet this legislation, aluminum top producers have introduced the button down, push-button end, and the flip top, nondetachable end. Instead of reverting back to the original flat top end which requires a can opener, the beverage packaging industry catered to the consumer preference for convenience by developing a self-opening, nondetachable end. The flip top, nondetachable end, although somewhat more expensive,

proved to be the preferred end. The button down end is more difficult to open. Thus, consumers apparently are willing to pay more in terms of costs and material use for convenience.

HEALTH AND SAFETY AND CONSUMER PROTECTION. Before a container can be used to package beverages and sold commercially, government approval requires a waiting period during which specific tests must be conducted. This produces a lag between the development of a new container and its entry into the market. During World War II and the Korean War, this legislation was bypassed and a number of extremely fragile (glass) containers were manufactured. Under more normal conditions, these bottles would not receive approval. Other government regulations include the labeling of saccharin in diet soft drinks (which caused a decline in the consumption of these beverages), the banning of the 10-ounce soft drink can in the mid-1950s (because it looked too much like the 12-ounce can), and the banning of acrylonitrile copolymer (AN) resins because of their association with cancer in laboratory animals.

Market Structure and Conduct

The market structure and conduct of the beverage container industry have also influenced the apparent determinants of tin use. This section examines three important aspects of market structure and conduct—plant size, vertical integration, and the competition among producers of the different types of containers.

PLANT SIZE. The average beer brewer is considerably larger than the average soft drink bottler. Table 2-11, for example, shows that in 1977 an average plant produced 1,598 barrels of output in the beer industry, but only 113 thousand barrels in the soft drink industry. In earlier years,

Table 2-11. Beverage Production, Number of Plants, and Average Plant Size

	1950	1960	1970	1977
Soft Drinks				
Barrelage output (millions of barrels)	48.5	71.4	149.8	235.6
Number of plants	6383	4519	3054	2084
Output per plant (thousands of barrels)	7.6	15.8	49.1	113.0
Beer				
Barrelage output (millions of barrels)	83.5	88.9	122.6	155.0
Number of plants	407	200	154	97
Output per plant (thousands of barrels)	205.2	444.5	796.1	1597.9

Sources: U.S. Brewers Association (various issues), National Soft Drink Association (various issues), Demler (1980).

the relative size discrepancy between the two was even larger.

Earlier sections have documented the considerable effect that new containers and other technological developments have had on tin consumption in the beverage industry. These sections have also indicated that the beer industry has consistently been quicker to introduce and adopt these new technological innovations. Only the plastic bottle, which recently appeared in the soft drink market, is an exception.

Part of this difference in behavior between the beer and soft drink industries is due to differences in plant size. The more disaggregated structure of the soft drink industry has historically encouraged can and bottle manufacturers to market their new products first to brewers. The larger size of brewers allowed them to spread the high capital cost of the new filling equipment over a greater volume. It also helped ensure that minimum order sizes required by container manufacturers, which are some three times greater for cans, were achieved. Finally, brewers served wider geographic markets, and so could realize greater savings in introducing one-way bottles and cans and thereby eliminate the reshipping charges incurred with the traditional returnable bottle.

Over the period, the average size of brewer and soft drink bottler has grown significantly—about 8-fold for brewers and 15-fold for bottlers. The relative higher rate of increase for soft drink bottlers appears to have had an impact on the timing of container commercialization. Table 2-12 indicates the years containers were first used commercially in the beer and soft drink industries. It shows that the lag between the beer and soft drink industries has declined considerably over time, from eighteen years for the three-piece tinplate can to less than two years for the two-piece tinplate can. A very recent innovation is the "stay-on" top, which was commercially introduced in both the beer and soft drink industries in the same year (1978). Thus, the growth in firm size found in the soft drink bottling industry, which followed the lead set by the brewing industry, apparently has accelerated the adoption of new innovations. This in turn has affected the use of tin and other materials in beverage containers.

VERTICAL INTEGRATION.   Over the 1950–77 period, there was a trend toward greater vertical integration in the beverage container industries. During the 1940s and 1950s, sheetmaking was the domain of steelmakers, can fabrication was the job of the canmaking firms, and beverage producers purchased containers from either canmakers or bottle manufacturers. In 1958, Hawaii Brewing, and then Coors Brewing, fabricated the first aluminum beer cans. This backward integration of beverage producers into canmaking represented a significant departure from historical experience. Later Kaiser Aluminum entered the canmaking business, fabricating an economic aluminum can in 1963. For the first time, beverage can sheet producers had integrated forward

Table 2-12. Introduction of New Beverage Containers (Year Commercial Innovation Introduced)

Innovation	Beer	Soft Drinks	Years Difference
Metal Cans	1935	1953	18
One-Way Bottles	1935	1948	13
Aluminum Cans	1958	1967	9
Aluminum Tops	1961	1966	5
Tin-Free Steel	1967	1969	2
Two-Piece Tinplate	1971	1973	2

Source: Demler (1980).

into canmaking. By 1968, five beverage packagers were manufacturing their own cans, including Coors Brewing, Anheuser Busch, Miller Brewing, Coca Cola, and Pepsi Cola.

The move to captive container production facilities favored the can in its competitive struggle for the packaged market. The economies of scale and the much more sophisticated technology involved in producing glass containers forced most beverage producers who were interested in manufacturing their own containers to use cans.

Another important aspect of market structure related to vertical integration concerns the close relationship between can manufacturers and the steel firms producing tinplate that existed from the entry of the three-piece tinplate can until at least the mid-1960s. As a result of these ties, can manufacturers during the 1950s and early 1960s shied away from experimenting with aluminum cans. It was largely for this reason that the aluminum can was first introduced by beverage producers and then by aluminum companies— Kaiser and Reynolds Aluminum.

COMPETITION AMONG CONTAINERS.   The development, introduction, and diffusion of new containers and other technological developments have been stimulated by intense competition among the producers of the different types of containers. This is illustrated by table 2-13, which identifies the major innovations occurring in the beverage container industry over the 1935–77 period.

The three-piece tinplate can first appeared in 1935, and that very year the glass industry introduced a one-way container. The glass and the steel industry experimented with their new containers over the next twenty years in direct competition with each other. In 1959, the glass industry introduced the glass-can, which was an effective substitute for the metal can. In that same period, Coors and Hawaii Brewing introduced the aluminum can. In 1961, the steel canmaking industry started to fabricate steel cans from cheaper double-reduced steel, which helped forestall market penetration from glass and aluminum.

In 1963, Reynolds Aluminum entered the beverage market with their version of the two-piece aluminum can. It wanted to gain a firm foothold in the beverage market, and so offered the container at a price competitive with steel cans. With a portion of the beverage industry converting to aluminum, the steel and canmaking industry introduced the tin-free steel can, and offered it at a discount to both the tinplate and aluminum can. The aluminum industry, threatened by this development, responded by producing cans from its efficient H19 aluminum alloy. This saved its competitive advantage, and allowed it to continue to expand its market share.

In 1971, the steel industry in turn responded by introducing the two-piece steel can, and thus was again able to compete with aluminum. During the 1970s, the steel and aluminum industries also effected weight and cost reductions to increase their share of the beverage market.

The plastics industry entered the beverage container market with AN-resins during the late 1960s. Glass attempted to block this competition with plastic coatings on glass con-

tainers. Later, glass began to utilize the press-and-blow technique, which directly competes with cans and plastics. In 1976, AN-resins were banned (because of their association with cancer), but the plastics industry almost immediately commercialized the "on-shelf" technology of PET-resins in their attempt to maintain and increase their market share.

In reviewing the technological history of the beverage container industry, one finds that many innovations were stimulated by competition between materials and container types. The commercialization of new innovations, which had been in the development stage for some time, apparently was prompted by the pending loss of market share to competitive containers.

There is, however, a risk associated with introducing new containers, particularly for beverage producers. Containers

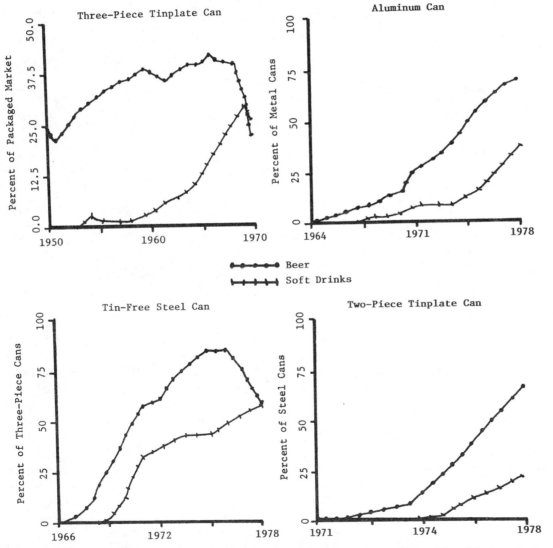

Figure 2–12. Commercialization of new containers into the beverage market. [From *Modern Packaging* (various issues); *Modern Metals* (various issues); *Tin International* (various issues); Way (1979); U.S. Brewers Association (annual); and National Soft Drink Association (annual).]

Table 2-13. Major Innovations in the Beverage Container Industry

Year of Commercialization	Nature of the Innovation and Sector of Beverage Container Industry in Which It Occurred				
	Glass	Plastic	Steel	Aluminum	Comment
Before 1935	Returnable bottle				First beverage container
1935 (early in)			Three-piece hot-dipped tinplate can		Steel and canmaking industry penetrate the beverage market
1935 (later in)	One-way glass bottle				Glass industry introduces a one-way container to compete with can
1935 to 1958	Glass container light-weighting				Competition to regain lost market share to metal can
1946 to mid-1950			Hot-dipped tinplate to differentially coated tinplate to .25 lb. electrolytic tinplate		World War II and Korean War and economics are the determining factors
1958				First aluminum beverage can	Manufactured by Hawaii Brewery and Coors Brewery
1959	"Handy" glass containers				Glass industry's competitive container to metal can (weight efficiencies, strength, cost)
1961			Double-reduced steel		Steel and canmaking industry switch beverage can over to cost-efficient DR steel
1963				Reynolds Aluminum two-piece can	Aluminum industry enters beverage container market and increases market share
1965				Reynolds Aluminum "necked-in" can	Improves container & competes more effectively with DR steel

that fail to meet shelf-life specifications, that alter the taste of the beverage, that suffer deterioration in appearance under various environmental conditions, or that for other reasons fail to gain consumer acceptance can destroy a beverage producer's market image within weeks. This aspect of the competition among brewers and soft drink bottlers delays the widespread adoption of most new containers for several years. As shown in figure 2-12, there typically is a period of slow growth after the commercialization of new beverage containers.

Tin Consumption in Beverage Containers

In 1950, the United States used some 6,300 tons of tin to manufacture beverage containers, all of which was used for beer cans since at that time the glass bottle monopolized the soft drink container market. Tin consumption declined during most of the 1950s, and then partially recovered dur-

ing the 1960s as the tinplate can penetrated the soft drink market. In the latter part of the 1960s, however, a new decline set in, and by 1977 the country was using only about 1,600 tons of tin in its beverage containers.

This chapter examined the causes for this erratic though dramatically downward trend in the consumption of tin on two levels. On the first level, that concerned with the apparent determinants, by far the most important development causing the decline in tin use was the reduction in the tin content of the tinplate can. The negative influence of this apparent determinant was reinforced in the case of beer by a decline in the share of the packaged container market held by the tinplate can, and in the case of soft drinks by a rise in bulk containers. On the positive side, a substantial increase in beer and soft drink consumption tended to increase tin use over the period. Its effect was reinforced in the case of beer by the relative decline in bulk containers and in the case of soft drinks by the increase in the share of the pack-

Table 2-13. Continued

| Year of Commercialization | Nature of the Innovation and Sector of Beverage Container Industry in Which It Occurred | | | | |
	Glass	Plastic	Steel	Aluminum	Comment
1967			Tin-free steel can		Steel and canmaking industries' attempt to maintain beverage market
1969				Aluminum can from H19 Alloy	Aluminum industry competes with TFS and increases market share
Late 1960–1976		AN-Resins			Plastics enter the beverage container market
1970	Plastic coatings				Glass industry maintains market share by threat from plastic containers
1971			Two-piece tinplate cans		Steel competes directly with aluminum can and attempts to regain lost market
1972–1974				H19 alloy more efficiently utilized	Aluminum industry in direct competition with steel
1974	Kerr-Heye glass				Press-and-blow technique, lighter and stronger glass container to compete with cans and plastic
1976			Miraform II	Miraform II	New bottom-profile can which both steel and aluminum use to compete with each other
1976		PET containers			After ban on AN resins, plastics industry attempts to gain market share with PET resins

Source: Demler (1980).

aged market acquired by the tinplate can. The average size of the tinplate can used for beer and soft drinks changed very little over time, and so had no significant effect on tin consumption.

On the second level, the underlying factors responsible for these changes in apparent determinants were identified and assessed. The rise in beverage consumption, along with changes in the relative importance of bulk and packaged containers, occurred primarily as a result of trends in population, per capita income, and consumer preferences. Technological change also appears to have had some influence, particularly by providing greater convenience and accessibility. In contrast, the prices of tin and the other materials used in beverage containers had no significant influence on these determinants. They represent simply too small a fraction of the final price to consumers of a beer or soft drink to affect their behavior.

The important decline in the tin content of tinplate cans and the fluctuating use of the tinplate can in the packaged market are largely explained by the many technological innovations that the dynamic container industry has produced over the last three decades. The size of firms, the intense competition between producers of different types of containers, the trend toward vertical integration, and material prices have all influenced the speed with which new products are developed, introduced, and adopted.

The price of tin has a largely indirect effect on tin consumption through the incentives and direction it provides for research and development. The nature of container technology has not allowed producers of tinplate cans to switch to tin-free steel, glass, or aluminum when the price of tin goes up. Nor can they reduce the tin content of the tinplate they are using by substituting some other material. In the future, producers may have the option of switching quickly and cheaply from tinplate to aluminum in response to changes in material prices if the dual-canmaking facilities introduced in 1976 are widely adopted.

3

Solder

Patrick D. Canavan

The class of metal alloys known as solders are material inputs in the production of numerous capital and consumer goods. Their unique properties enable the molten metal to readily flow onto and permanently adhere to (or wet) the surfaces of a number of common metals, such as copper and copper-base alloys, tin and tinplate, and lead. The narrowly separated surfaces of the components are thus joined by the film of solder.

The primary constituents of solder are tin and lead. Many desirable properties such as shear strength, wetting action, fluidity, and thermal and electrical conductivity increase in a nonlinear fashion with increasing tin content up to about 63 percent tin in terms of weight. Tin can be replaced with lead to save costs only at the expense of these properties. Additions of small amounts of silver or antimony mitigate somewhat the adverse effects of a reduction in tin content and augment certain desired properties, particularly strength.

Less costly and less strategically vulnerable substitutes for tin solder have long been sought, but as L. Parry (1905, p. 451) stated shortly after the turn of the century: ". . . solder [is] required on account of such special and definitive combination of physical properties that [tin's] replacement by any other metals to any appreciable extent is quite impossible." Even after two world wars and the attendant compulsory "tin conservation" efforts, the 1952 Paley Commission concluded that "tin-lead soft solders are so widely and easily used that substituting other materials is probably more difficult than for any other general application of tin" (U.S., President's Materials Policy Commission, 1952, vol. IV, p. 60).

The difficulty of substituting away from tin is at first appearance borne out by figures showing the average composition of solder produced in the United States since 1950.

As figure 3-1 indicates, the aggregate tin content of solder has remained quite static, varying between 21 and 26 weight percent tin since 1950, with only the years 1973 and 1974 falling somewhat outside this range. This constancy of composition has been maintained even though the ratio of the price of tin to lead has increased sharply, from 7.5 in 1950 to 18.7 in 1978. With such price incentives to reduce the use of tin, this constancy of composition suggests that the substitution of other materials for tin in solder is indeed infeasible.

However, subsequent sections will show the apparent constancy of post-1950 average tin content to be, in part, a product of aggregation. Reductions in the tin content in some solder applications, such as radiators and evaporated milk can seams, have been offset by the growth of the electronics industry, which utilizes 60-63 percent tin solder almost exclusively.

Moreover, looking back before 1950, one finds much less support for the premise of insubstitutability even in the aggregate data. During World War II, the average tin content of solder was cut in half, with the bulk of the reduction occurring within one year, 1942. Experts maintained through the midst of the war-induced tin conservation program that reduction of the tin content in solder was achievable only by the "willingness of users, on patriotic grounds, to put up with more or less inconvenience" (Ireland, 1943, p. 10). However, the average tin content of solder did not rebound to prewar levels after all governmental restrictions on tin usage were removed at the conclusion of the Korean War.

Figure 3-2 shows that while annual solder output levels have been quite volatile, the trend has been stagnant in the postwar period. Since the U.S. economy has expanded greatly since World War II, the intensity of solder use with respect

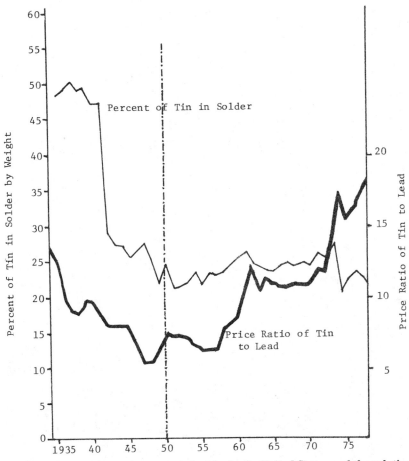

Figure 3–1.   Average tin content of solder in the United States and the relative price of tin to lead, 1935–78. [From American Metal Market (various years), and U.S. Bureau of Mines (various years).]

to industrial output has declined persistently and markedly. As depicted by figure 3-3, the consumption of solder and tin in solder per unit of industrial production has declined by three-fourths since its peak in 1950.

Some of this decline conceivably could have resulted from faster growth in industries that use little or no solder. However, this appears unlikely, for the use of solder is concentrated in economic sectors such as metal containers, transportation equipment, electrical and electronics equipment, and construction that have enjoyed relatively rapid growth over the postwar period. As demonstrated in subsequent sections, the decline in intensity of use can be largely explained by the adoption within traditional solder-consuming industries of new product designs and production techniques that require less solder per unit of output or eliminate the use of solder completely. This innovative activity is, in part, the result of the high and rising price of tin, but other factors are important as well.

This chapter examines quantitatively the consumption of the tin in solder used for three major applications—seams for metal cans, motor vehicle radiators, and new motor

vehicle bodies. For each of these end uses, the apparent determinants of tin usage are identified and assessed, along with the underlying factors responsible for changes in the apparent determinants. In a more qualitative manner, the chapter also considers tin and solder consumption in electronic and plumbing applications.

Table 3-1 shows the 1978 consumption of solder in the United States by end use industry. It also indicates the amount of solder used in the three applications examined in greatest depth in this chapter. While these applications account for nearly 45 percent of total 1978 U.S. solder consumption, the tin contained therein accounts for less than 17 percent of the tin contained in all solder. This is a function of the low tin content of the solders used for can seams and automobile body work. In contrast, electrical and electronic appliances require solders with particularly high tin content.

This chapter focuses primarily on solder usage in the United States since 1950. Some comparisons are made with trends in Great Britain. Also, the use of solder before World War II is considered briefly. Since World War II stimulated intensive adaptive activity to offset shortages and govern-

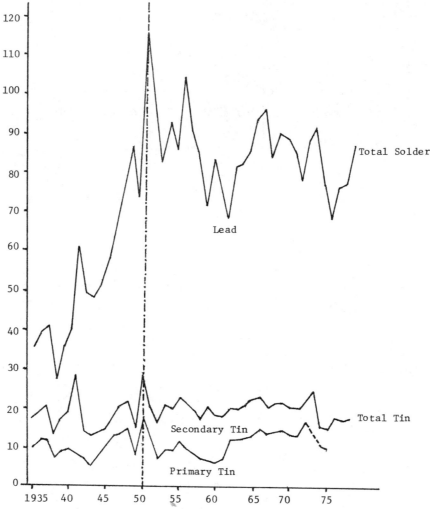

Figure 3–2.   U.S. consumption of tin and lead for solder, 1935–78. [From U.S. Bureau of Mines (various years), and American Metal Market (various years).]

ment restrictions on strategic materials, including tin solders, this era of compelled substitution is examined in somewhat more detail.

Solder Usage in Metal Cans

Solder has been used to make hermetic seams for three-piece tinplate containers since the innovation in the early nineteenth century of food canning technology. Prior to the advent of the sanitary can in the first decade of this century, solder was used to seal the top and bottom can closures and the side seam of the cylindrical can body. The contemporary evaporated milk can is a remnant of this all-soldered construction and will be discussed below. The far more common sanitary can has a soldered interlocking side seam joint which closes the can body, but can top and bottom closures

are crimped on and hermetically sealed with organic cements.[1]

Two variants of the soldered lock seam are illustrated in figure 3-4. The narrow clearance between the tinplate hooks is infilled by capillary flow when the molten solder contacts the outer portion of the seam. For most fruit and vegetable containers, solder infilling is specified for the outermost two-thirds of the joint clearance, as shown in figure 3-4a. This type of joint is a partial fillet. Cans used for aerosols, beverages, and some high-acid products require solder penetration into the interior portion of the seam (figure 3-4b). The complete infilling of all joint clearances (or a full solder

---

[1]The use of the word "sanitary" for this type of can construction does not imply that the all-soldered, evaporated milk can or other can types are unhealthful. Rather, the first producers of the can, the Ams Company, chose to promote the solderless sealing of can closures as a sanitary measure, and the term "sanitary can" has since referred to all tinplate cans that have closures sealed with organic cements.

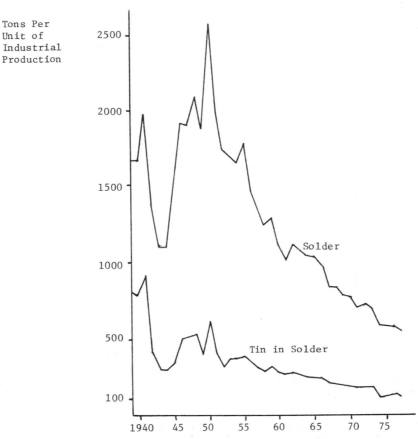

Tons Per
Unit of
Industrial
Production

Figure 3–3.   Intensity of solder and tin in solder consumption: tons per unit of the Federal Reserve Board's index of industrial production (1967 = 100 units). [Units of industrial production are from the Board of Governors of the Federal Reserve System as reported by U.S. Council of Economics, 1980. Data on the consumption of solder and tin in solder are from U.S. Bureau of Mines (various years), and *American Metal Market* (various years).]

fillet) provides a stronger joint that is capable of withstanding greater pressures and a smoothed interior seam that covers the exposed metal edge, which could rupture the internal coatings.

With the exceptions noted below, sanitary cans have had side seams sealed with 2 percent tin-98 percent lead (2/98) solder since 1950. The relationship between the can height and the pounds of 2/98 solder required per 1,000 cans, estimated in figure 3-5, has remained basically unchanged since at least the early 1960s. Actual usage rates may vary about the values shown because the size of the solder fillet may be tailored to specific product requirements, and each can manufacture has its own design specifications. However, since more detailed data are lacking, the rates given in figure 3-5 are used along with information on the number of cans produced and their average size to estimate tin consumption in can seam solder since 1950.

The Can Manufacturers Institute and the U.S. Census Bureau both estimate the annual number of metal cans shipped, broken down into some 18 product classes. This section evaluates solder usage trends for only four of these 18 product classes: fruits and vegetables, beer and soft drinks, aerosols, and evaporated milk. In 1950, these four product classes accounted for 57 percent of all metal cans shipped. By 1978, their share of the total increased to nearly 80 percent, primarily due to the growth of the canned beverages sector, which now comprises over 50 percent of all metal can shipments.

Figure 3-6 summarizes the postwar trends in tin consumption for can seam solder for each of these four product classes. Analysis of these trends follows a brief description of can solder usage before 1950.

Can Solder Before and During World War II

The technology of three-piece tinplate can fabrication has not changed markedly since before World War II, although the composition of solders in use has. Hiers, in a 1931 article, stated that the common canmakers' solders in use were 40/60 and 50/50. H. W. Phelps, then president of

Table 3-1. U.S. Consumption of Solder and Tin in Solder by End Use Industries, 1978

End-Use Industry	Tin in Solder (tons)	Solder (tons)
Metal Cans	890–1,420	20,745–21,285
4 product types (F&V, beverage, aerosol, milk)[a]	487	10,179
Motor Vehicles	4,445–5,330	22,845–23,740
New car and light truck radiators[b]	2,136	13,349
New car bodies[b]	422	14,346
Electrical and Electronics	8,000–8,885	18,440–19,340
Coatings	890–1,420	5,635–10,310
Plumbing and Other	715–3,545	11,485–18,495
Total	17,770	86,160

Sources: Data for the major end use industries are estimates based on information obtained from the U.S. Bureau of Mines. The figures for total are also from the U.S. Bureau of Mines. Data for new car radiators, new car bodies, and F&V, beverage, aerosol, and milk cans are estimates based upon the apparent determinants of usage as explained in the text.

[a]F&V are fruit and vegetable cans.

[b]Estimate of new car radiator plus new car body solder exceeds the total motor vehicle estimate partly because the new car radiator figure includes home scrap solder, which is reclaimed by automakers for use as body solder.

American Can Company, stated before the 1934 Tin Investigation that 50/50 solder was always used to seal the seam of processed food cans, suggesting that the lower tin content solders were used in nonfood cans. By 1941, it appears that the average tin content of can solder had fallen slightly from this level. Lueck (1942, p. 62) states that the most prevalent can solder in use immediately prior to World War II was 40/60, while Hartwell (1952, p. 57) maintains that the "customary" prewar solder contained 37 percent tin.

Many months before the entry of the United States into World War II, strategies to cope with the impending shortage of tin were outlined. Decreasing the amount of tin in solder was identified as a principal means of conserving tin (National Academy of Sciences, 1941, p. 45). Within ten days of the attack at Pearl Harbor, the War Production Board issued Order M-43, which restricted the tin content in all solders to 30 percent. The allowable tin content was lowered successively, down to 21 percent by early 1944.

According to an Army Industrial College report (1945, p. 21), the small can manufacturers found it difficult to comply with the mandated tin content reductions. On the other hand, the three major can manufacturers responded by demonstrating the applicability of substitute solders to

A. Partial fillet   B. Full fillet

Figure 3–4.   Soldered container side seam: cross-sectional view.

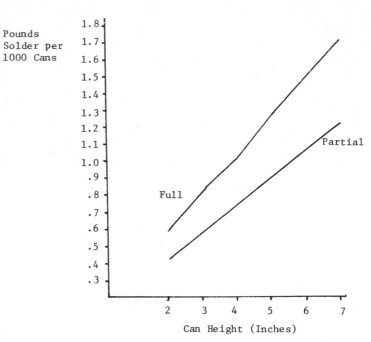

Pounds Solder per 1000 Cans

Figure 3–5.   Usage of 2 percent tin–98 percent lead solder for full and partial side seam fillet versus can height. [From Continental Can.]

high-speed automated canmaking. The properties of silver-lead solders were well known, having been used for years by Westinghouse Electric (U.S. War Metallurgy Committee, 1943, p. 420), and the U.S. Treasury held nearly 100,000 tons of silver. Thus, silver-lead solders were the focus of industry substitution efforts well before the war began (National Academy of Sciences, 1941, p. 21).

During the conversion of the existing can lines to the use of lead-silver solder with a higher melting temperature, it was found that the presence of 3-5 percent of tin in the molten lead-silver bath improved solderability. With hot dip tinplate, tin from the hot dip coating would dissolve into the solder bath, yielding this desired tin level, or even more, requiring the addition of pure lead to dilute the tin. However, during this same period, electrolytic tinplate was rapidly being introduced. As less tin from the electrolytic plate dissolved into solder bath, 2-3 percent tin was added to the lead-silver alloy. Lueck and Brighton (1944, p. 534) found that low tin solder on electrolytic plate resulted in a stronger seam than traditional 40/60 solder on hot dip tinplate.

Toward the end of the war, can manufacturers first reduced the silver content of the substitute solder to 1 percent, and then eliminated silver altogether, without significantly detracting from the performance of the joint (Van Vleet, 1948, p. 320). By the end of the war, major manufacturers had thus adopted the use of 3/97 solder, which with tin pick-up from electrolytic tinplate yielded a stabilized bath composition of 5 percent tin. By the Korean War, the tin content of can solder had been lowered further to 2 percent.

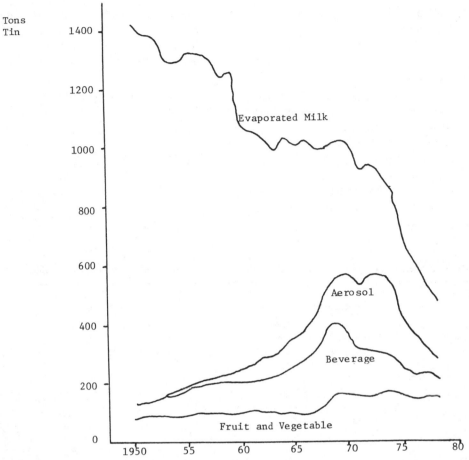

Figure 3–6. **Tin in solder consumption for fruit and vegetable, beer and soft drink, aerosol, and evaporated milk containers, 1950–78.** [From tables 3–2, 3–3, 3–4 and 3–5.]

During World War II, an intensive effort was also made to develop hermetic welded and cemented three-piece can side seams. While unsuccessful at the time, this research did eventually promote the commercialization of such containers in the mid-1960s.

The Apparent Determinants of Tin Consumption

The amount of tin used for any particular solder end use, such as fruit and vegetable can seams, beverage can seams, or automobile radiators, during year t depends on four apparent determinants, as indicated by the following identity:

$$Q_t = a_t \, b_t \, c_t \, X_t$$

where: Q_t = the tons of tin used in a particular solder application during year t

a_t = the average tin content (by weight proportion) of the solder used for that end product during year t

b_t = the weight amount of solder used per unit of soldered end product during year t

c_t = the proportion of soldered products in the total end product population during year t

X_t = the number of end products (cans, radiators, autobodies, etc.) fabricated in year t

In this chapter, the number of end products is considered exogenous and little effort is made to explain changes in this variable. The focus, instead, is on the three other apparent determinants and the underlying factors causing them to change.

In the previous section the decline in the apparent determinant a_t (average tin content), resulting from substitution of tinless and low tin solder for traditional 40/60 canmakers solder, was found to be the single most important development responsible for the decline in the amont of tin used for can seams during World War II. The underlying factor causing the change in a_t was the necessity of a rapid response to emergency material shortages and government allocative control of the tin supply. These emergency conditions led to the development and adoption of the new solder and tinplates that yielded equivalent or superior service with less tin input. One can only speculate as to when these technological advances would have been commercialized had World War II not spurred their use.

Fruit and Vegetable Cans

Table 3-2 presents the apparent determinants of tin consumption for fruit and vegetable (F&V) can manufacture along with the resultant total number of tons of solder and tin in solder consumed annually for 1941 and the years 1950 through 1978. Figure 3-7 shows how the quantity of tin in solder per million cans has changed since 1950.

Over the 1950–67 period, according to table 3-2, the use of tin in F&V can side seams varied in direct proportion with the number of cans produced, as the three other ap-

parent determinants of tin consumption remained unchanged. It is for this reason that the intensity of tin use for F&V containers, shown in figure 3-7, is constant at 16.4 pounds per million cans over this period.[2]

[2]The figure of 16.4 pounds per million is based upon the deduction that all fruit and vegetable containers produced during this period (a) were three-piece soldered tinplate cans, (b) averaged 4½ inches in height, and (c) had partially filleted side seams filled with two percent solder. Conditions (a) and (b) imply that .82 pound of solder was required per thousand cans (figure 3-5), which times .02 pound of tin per pound of solder gives .0164 pound of tin per thousand cans, or 16.4 pounds per million cans. While the National Canners Association identifies 22 container sizes commonly used in canning fruits, vegetables, and juices, ranging in height from 2³⁄₁₆ to 7 inches, three cans have dominated the container selection, the No. 300 can at 4⁷⁄₁₆ inches, the No. 303 at 4⁶⁄₁₆ inches, and the No. 2 can at 4¹¹⁄₁₆ inches. (continued on p. 43)

Table 3-2. Consumption of Solder and Tin in Solder for Fruit and Vegetable (F&V) Cans, 1941, 1950–1978

			Conventional Fillet				High Tin Fillet				
Year	Total F&V Can Production (billions)	Proportion of F&V Cans with Soldered Seams	Can Production (billions)	Solder Consumed per 1,000 Soldered Cans (pounds)	Tin Content of Solder (pounds per pound solder)	Tin Consumed (tons)	Can Production (billions)	Solder Consumed per 1,000 Soldered Cans[a] (pounds)	Tin Content of Solder (pounds per pound solder)	Tin Consumed (tons)	Total Tin Consumed in F&V Can Solder (tons)
1941	9.10	1.00	9.10	1.00	.40	1,651					1,651
1950	10.84	1.00	10.84	.82	.02	81					81
1951	12.55	1.00	12.55	.82	.02	93					93
1952	12.42	1.00	12.42	.82	.02	92					92
1953	12.84	1.00	12.84	.82	.02	96					96
1954	12.44	1.00	12.44	.82	.02	93					93
1955	13.85	1.00	13.85	.82	.02	102					102
1956	15.11	1.00	15.11	.82	.02	113					113
1957	13.84	1.00	13.84	.82	.02	103					103
1958	14.45	1.00	14.45	.82	.02	108					108
1959	14.34	1.00	14.34	.82	.02	107					107
1960	14.26	1.00	14.26	.82	.02	106					106
1961	14.25	1.00	14.25	.82	.02	106					106
1962	15.56	1.00	15.56	.82	.02	116					116
1963	13.82	1.00	13.82	.82	.02	103					103
1964	14.27	1.00	14.27	.82	.02	106					106
1965	14.35	1.00	14.35	.82	.02	107					107
1966	12.74	1.00	12.74	.82	.02	95					95
1967	14.51	1.00	14.51	.82	.02	108					108
1968	15.52	1.00	15.45	.82	.02	114	.07	.94	.995	29	143
1969	16.10	1.00	15.97	.82	.02	118	.13	.93	.995	54	172
1970	15.60	1.00	15.49	.82	.02	115	.11	.93	.995	48	163
1971	15.30	1.00	15.19	.82	.02	112	.11	.94	.995	46	158
1972	15.00	1.00	14.09	.82	.02	110	.11	.95	.995	49	159
1973	16.30	1.00	16.19	.82	.02	120	.11	.94	.995	49	169
1974	16.80	.99	16.69	.82	.02	123	.11	.95	.995	48	171
1975	16.00	.99	15.91	.82	.02	116	.09	1.04	.995	43	159
1976	14.67	.98	14.58	.82	.02	105	.09	1.04	.995	43	148
1977	15.22	.97	15.11	.82	.02	108	.11	1.08	.995	52	160
1978	15.24	.96	15.14	.82	.02	109	.10	1.05	.995	47	156

Sources: Data on F&V can production are from the Can Manufacturers Institute and the U.S. Bureau of Census. Estimates for conventional fillet solder usage per 1,000 cans, 1950–1978, are from Continental Can (see figure 3-5). All other data are estimated, as explained in the text.

[a]The varying coefficient for HTF solder consumed per 1,000 HTF soldered cans reflects the shifting proportions of HTF canned tomato packs, which at 4⁶⁄₁₆ inches in height utilize 1.4 pounds per 1,000, to HTF canned asparagus packs, which at 7 inches in height utilize .9 pound per 1,000, rather than any technical change in usage per standard can size.

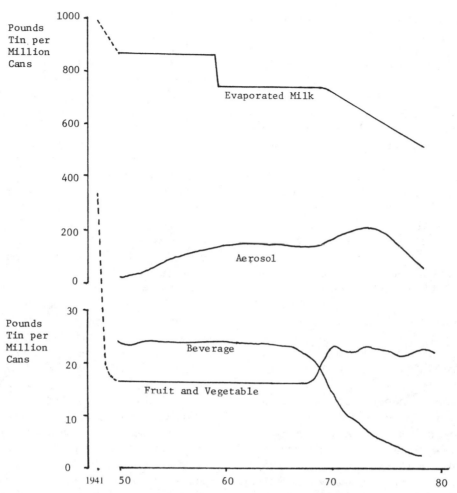

Figure 3–7. Intensity of tin in solder consumption for evaporated milk, aerosol, beverage, and fruit and vegetable containers, 1941, 1950–78 (pounds per million cans). [From tables 3–2, 3–3, 3–4 and 3–5.]

Although the tin content of solder used to produce F&V cans with conventional fillets has remained at 2 percent, the average tin content of the solder used for all F&V cans increased after 1967 with the adoption of the high-tin-fillet (HTF) can. This can, patented by Gilbert Kamm of the American Can Company in 1966, is basically a three-piece tinplate container that utilizes a full fillet of nearly pure tin solder. It was designed for packing relatively corrosive food products that rapidly attack tinplate, resulting in an unappetizing can interior or food. Product quality is thus improved by substituting pure tin for conventional 2 percent tin solder.

While initially considered for at least six different product packs (*Tin International*, 1966, p. 270; Hotchner and Kamm, 1967), the high tin fillet has been used commercially for only a portion of the asparagus and concentrated tomato product packs.[3] As table 3-2 indicates, HTF usage has required some 40-55 tons of tin annually since 1969, or about one-third to one-half of the tin in solder utilized for all other fruit and vegetable containers.

In addition to raising the average tin content of solder, the adoption of the HTF can also increased the average amount of solder consumed per thousand cans, as it has a full fillet side seam. Indeed, the increase would have been even larger were it not for the fact that HTF cans, particularly those used for tomato packs, tend to be relatively large compared with other F&V cans, which reduces the ratio of the side seam length per unit volume.

[3]More precisely, it is estimated that since 1969, three-fourths of all asparagus packs, and 10 percent of all No. 10 tomato product cans, increasing to 30 percent after 1975, have utilized the high-tin-fillet. The increase in HTF usage for tomato packs after 1975 was due to concerns over the concentration of lead in the highly acidic pack. Tin consumption is about 900 pounds per million 4⁹⁄₁₆ inch high asparagus cans and 1,400 pounds per million 7 inch high No. 10 tomato cans.

The average height of these three cans is 4½ inches. The data do not indicate any pronounced post-1950 change in average can height.

The proportion of soldered fruit and vegetable cans to all fruit and vegetable cans declined slightly following the introduction of two types of seamless two-piece containers and a three-piece welded container during the latter half of the 1970s. Previously, virtually every fruit and vegetable can had been a three-piece soldered tinplate container. The new containers may eventually completely replace the soldered can but, as shown in table 3-2, they had made only modest inroads into the fruit and vegetable container market by the end of the 1970s.

The two-piece D&I (drawn-and-ironed) can, initially introduced for the beverage can market, has been used since 1976 by Campbell's for dog food and V-8 vegetable juice (Melville, 1976, p. 13). Two years later, Del Monte was making D&I cans for vegetable packs (Ewart, 1978, p. 23). Since an annual volume of 200-250 million cans is necessary to amortize the five-million-dollar investment in a D&I line, future penetration of D&I fruit and vegetable cans will concentrate on very high volume items that can support such a large output of identical cans. Furthermore, economic utilization of the D&I technology is limited to cans in which substantial wall gauge thinning is possible.[4] While the internal pressures exerted by the contents of aerosol and beverage cans make the D&I can stronger after filling, the internal vacuum resulting from processed food retorting and cooling tends to collapse the thin walls of the D&I can.

Most experts predict that the seamless draw-redraw (D&RD) container will eventually have wider application in food containers than the D&I container. The capital cost of a D&RD line is 40 percent less than a D&I line, the D&RD line can be readily retooled to make different can sizes, and it produces a can with stronger sidewalls. While the fabrication of large cans up to $4\frac{1}{16}$ inches in diameter by $4\frac{11}{16}$ inches in height is possible with D&RD technology, soldered cans currently have an economic advantage at sizes above $2\frac{11}{16}$ inch diameter by 3 inches in height because D&RD cans greater than this size require two or three draws, which greatly slows the output rate, increasing per-unit costs. As of 1978, the production capacity to make over 2 billion D&RD food cans annually was on stream. Of the dozen D&RD food lines, most were committed to production of shallow containers for tuna fish, beef stew, and pet food. Only one line is operated by a vegetable packer, Green Giant.

The Soudronic[5] three-piece welded tinplate container was rapidly adopted in the nonfood container market in the United States during the 1970s. In 1978, a Soudronic-compatible side seam stripping system was commercialized that extends the potential market of the welded container to food cans as well (Leimgruber, 1978, p. 18). Ewart (1978, p. 20) reports that the welded tinplate container is already in use for fruit juices and vegetables, but does not mention the employing firms or output figures. The Soudronic system is similar to a conventional soldering line in that a Soudronic canmaking line can quickly be retooled to accommodate a wide range of can sizes and costs about the same, five hundred thousand dollars.

As of 1980, these nonsoldered alternatives combined accounted for 3-5 percent of the total fruit and vegetable metal containers, and 10 percent of all food cans. It is likely that a conversion to nonsoldered containers will occur as new canmaking capacity is built to replace obsolete soldering lines. A major factor stimulating the obsolescence of soldered fruit and vegetable cans is the increasing public awareness over the health hazards of lead exposure.

The Food and Drug Administration (FDA) has been concerned with the potential danger of soldered can seams containing lead since the mid-1930s. For many years, it maintained that lead absorption by food packs from 40/60 and 2/98 solder posed no significant health hazard as long as good manufacturing practices were followed. This position came under scrutiny in the early 1970s with the aroused interest in heavy metal intake from all sources. It has been estimated that cans seamed with solder are responsible for 14 to 20 percent of the total dietary intake of lead by children and adults, up to 6 percent of total intake of lead from all sources (air, water, food) by adults, and up to 9 percent for children (*Federal Register*, 1979, p. 51233; Reinsch, 1979, p. I-1).

Public concern has been particularly directed toward infants and toddlers because they have a higher level of exposure per unit body weight and are more susceptible to lead toxicity than adults. In response to this concern, baby food packers first reduced lead dust formation on their soldering lines; then, convinced that regulatory action was imminent, they abandoned soldered containers altogether.

Action taken by the FDA on August 31, 1979, confirmed the expectations of baby food product packers, and sent a message to packers of adult foods as well. Reiterating policy goals stated 18 months earlier, the FDA (*Federal Register*, 1979, p. 51233) officially proposed to reduce by at least 50 percent within the next five years the contribution of lead intake from lead soldered cans. To achieve this, the FDA (*Federal Register*, 1979, p. 51240) proposed to establish lead content levels appropriate for designated products that take into account: "the extent to which the use of lead can be avoided and minimized in the manufacture process, the potential for disruption in the food supply, the level of risk posed, and the ubiquitous distribution of the substance."

---

[4]With the D&I production process, the sidewalls of the can are made to height by stretching the metal. With the most advanced technology, starting with can stock .012 inch thick the ironing process thins the stock to .004 inch. With this process, metal can be selectively distributed to the loci of greatest stress. With the D&RD process the can is punched from the sheet stock. Little wall thinning occurs and the metal cannot be selectively distributed.

[5]Soudronic S. A. is a Swiss company, and the Soudronic technology is substantially different from Continental Can's Conoweld process.

The crucial factor for the future of the soldered food container is the baseline from which the 50 percent reduction in lead is mandated. Lead from can solder enters the food pack three ways: migration directly from the solder in the side seam interlock and laps;[6] leaching from or ingestion of solder drops splattered onto the can interior; and of most importance, entrapped solder dust. FDA studies of the lead content of food packs before and after soldering lines were modified to control for solder dust estimate that prior to such line modification, over 80 percent of the contained lead came from solder dust contamination during can formation, and only about 10 percent came from lead migration from the side seam itself. If food pack lead contents circa 1974, that is, prior to industry-wide cleanup efforts, are utilized by the FDA as the baseline, the required reduction can be readily achieved at modest cost with line modifications that control the dust, perhaps combined with spraying the inside seam with organic coatings to reduce direct migration.

Federal regulatory limits have not been instituted, and are unlikely to be under the Reagan administration. However, the issue of lead toxicity has riveted the attention of can fabricators and food processors on the traditional sanitary fruit and vegetable container and the new, potentially superior alternatives to it, thus likely speeding the adoption of the nonsoldered alternatives.

Beverage Cans

Of the four apparent determinants of the consumption of tin in can solder, only two have varied since 1950 in the beverage container market: the demand for beverages packed in metal cans (X_t), and the proportion of cans that are soldered (c_t). Since chapter 2 examines both of these developments in some detail, the analysis here can be brief.

Table 3-3 presents the apparent determinants of tin consumption, along with the amount of total solder and tin in solder consumed in the fabrication of beverage containers annually over the 1950–78 period. Figure 3-7 shows the intensity of tin use in solder per million beverage cans. Since the amount of tin in solder required per million soldered cans has remained constant over the last three decades, the marked decline in intensity of tin use illustrated in figure 3-7 reflects the entry and market penetration of nonsoldered cans, which is directly documented in table 3-3.

Between 1963 and 1978, the market share of soldered beverage cans fell from 100 to 11 percent. Since the beverage can market was growing at an annual rate of 18 percent during 1964–68, the absolute number of soldered tinplate containers did not peak until 1968, four years after the two-

piece aluminum can and one year after the TFS can first appeared. For the reasons discussed in chapter 2, these new containers were introduced into the beer market two to three years earlier than into the soft drink market.

Among the various types of beverage cans in use, the three-piece soldered tinplate container has the highest metal cost. The price of solder, however, is not a significant factor in the switch away from the soldered can, since it accounts for less than 1.5 percent of the total metal costs. Indeed, more costly than the solder is the one square inch of tinplate per can that is utilized to form the side seam interlock. More specifically, for 1972 it has been estimated that the list price of the can stock for three-piece tinplate beverage containers was 14.55 dollars per thousand. Some 32 cents of this amount can be attributed to the tinplate used to form the lockseam. Meanwhile, solder costs ran about 22 cents per 1,000 cans. It has been similarly calculated that the elimination of solder is a significant but minor proportion (5 to 14 percent) of the total metal cost savings attributed to the adoption of the nonsoldered cans. Far more important is the reduction in cost of the can stock.

Improved can integrity has been an important factor in the adoption of nonsoldered cans. The soldered side seam is the weakest part of the container, whereas with the welded or cemented TFS container, the side seam is as strong or stronger than the body stock material. A two-piece container, of course, with its integral body and bottom, does not have a side seam and only one double-seamed closure. Both the soldered and cemented container have significant metal overlap at the side seam. This localized extra thickness in the cylindrical can body may result in an imperfect fit with the can closures. Often imperfections at this spot are a source of microleakage, resulting in spoilage of the perishable beverage product.

In general, the size and rapid growth of the beverage can market has provided strong incentives for canmakers to advance new and better containers. In these circumstances, the soldered can has been abandoned by the major beverage canners in favor of other types of containers that offer equivalent or superior service with the input of smaller amounts of less costly can stock.

Aerosol Containers

The historical trend of tin in solder usage in the fabrication of aerosol containers is somewhat more complicated than that of the other types of cans studied.[7] As table 3-4 shows, tin consumption in this end use grew from under one ton in 1950 to over 250 tons in the early 1970s, and then collapsed to less than 75 tons by 1978. While much of the

[6]The last quarter-inch of a can body's side seam on top and bottom is a simple overlap joint rather than an interlock, and is referred to as the side seam lap.

[7]Much of the information presented in this section was gathered from published and unpublished articles by, and conversation with, Montfort A. Johnson of Peterson/Puritan.

Table 3.3. Consumption of Solder and Tin in Solder for Beer and Soft Drink (Beverage) Cans, 1950–1978

			Beer			
Year	Can Production (billions)	Proportion with Soldered Seams	Solder Consumed per 1,000 Soldered Cans (pounds)	Solder Consumed (tons)	Tin Content of Solder (pounds per pound solder)	Tin Consumed (tons)
1950	5.0	1.00	1.2	2718	.02	54
1951	4.4	1.00	1.2	2385	.02	48
1952	5.0	1.00	1.2	2718	.02	54
1953	6.2	1.00	1.2	3401	.02	68
1954	6.6	1.00	1.2	3572	.02	71
1955	7.3	1.00	1.2	4000	.02	80
1956	7.8	1.00	1.2	4238	.02	85
1957	8.0	1.00	1.2	4372	.02	87
1958	8.3	1.00	1.2	4532	.02	91
1959	9.1	1.00	1.2	4977	.02	100
1960	8.8	1.00	1.2	4766	.02	95
1961	8.7	1.00	1.2	4723	.02	94
1962	8.9	1.00	1.2	4867	.02	97
1963	9.6	1.00	1.2	5250	.02	105
1964	10.7	.99	1.2	5789	.02	116
1965	11.3	.97	1.2	5934	.02	119
1966	12.8	.95	1.2	6582	.02	132
1967	13.6	.90	1.2	6627	.02	133
1968	16.4	.82	1.2	7317	.02	146
1969	18.1	.63	1.2	6165	.02	123
1970	19.5	.49	1.2	5196	.02	104
1971	20.2	.31	1.2	3416	.02	68
1972	21.8	.27	1.2	3143	.02	63
1973	24.1	.17	1.2	2185	.02	44
1974	26.1	.11	1.2	1541	.02	31
1975	26.1	.06	1.2	742	.02	15
1976	26.9	.04	1.2	551	.02	11
1977	27.9	.05	1.2	568	.02	11
1978	28.9	.05	1.2	560	.02	11

early growth in tin consumption can be attributed to the rise in aerosol can production, the intensity of tin use per million cans, shown in figure 3-7, has also varied—climbing in the 1950s, leveling off in the 1960s, rising again in the early 1970s, and then falling sharply in the mid- to late 1970s.

These fluctuations in intensity of use have occurred primarily as a result of (a) changes in the proportion of aerosol cans with soldered seams, and (b) changes in the mix of aerosol cans using low-tin and high-tin solder. Changes in the mix of low-tin and high-tin aerosol can solders have altered both the average tin content of solder and the average amount of solder consumed for the total aerosol can population, though these two apparent determinants have remained virtually unchanged for the subgroup of aerosol cans soldered with either a high-tin or low-tin solder.

Since the early 1950s, conventional 2 percent tin, 98 percent lead (2/98) solder has been utilized for about 70 percent of all soldered aerosol containers except for a short period in the early 1970s, as explained below. Very early

in the history of aerosol products, some types of containers soldered with 2 percent tin solder had a high incidence of side seam failure when subjected to prolonged periods of storage at elevated temperatures, such as in automobile glove compartments. The seam failure was not a sudden violent bursting, but rather a gradual unraveling of the seam, resulting in the nonviolent discharge of the contents. This phenomenon of solder creep can be mitigated by adding alloying agents, principally silver, but also in certain formulations copper, arsenic, antimony, mercury, and even tellurium are added (Kinnavy, 1965, p. 51). Within a few years, a solder developed at Continental Can, trade-named Tricom, and composed of tin (1.5 percent), silver (0.5 percent), and lead (98 percent), was in use for about 15 percent of all soldered aerosols, particularly automotive products.[8]

---

[8]At the peak of soldered aerosol container production, the early 1970s, about a ton of silver was consumed annually in Tricom or similar high-strength, low-tin solders for aerosol container side seams.

		Soft Drink				
Can Production (billions)	Proportion with Soldered Seams	Solder Consumed per 1000 Soldered Cans (pounds)	Solder Consumed (tons)	Tin Content of Solder (pounds per pound solder)	Tin Consumed (tons)	Total Tin Consumed (tons)
.1	1.00	1.2	40	.02	1	69
.5	1.00	1.2	258	.02	5	76
.3	1.00	1.2	180	.02	4	84
.3	1.00	1.2	167	.02	3	88
.4	1.00	1.2	200	.02	4	91
.4	1.00	1.2	217	.02	4	95
.5	1.00	1.2	290	.02	6	106
.8	1.00	1.2	435	.02	9	104
1.2	1.00	1.2	657	.02	13	107
1.6	1.00	1.2	885	.02	18	115
2.0	1.00	1.2	1104	.02	22	127
2.8	1.00	1.2	1499	.02	30	146
3.8	1.00	1.2	2078	.02	42	161
5.5	1.00	1.2	3017	.02	60	192
7.2	.99	1.2	3855	.02	77	210
9.9	.97	1.2	5214	.02	104	250
11.6	.97	1.2	6089	.02	122	245
12.9	.81	1.2	5698	.02	114	218
14.1	.61	1.2	4686	.02	94	162
15.6	.57	1.2	4834	.02	97	160
17.6	.51	1.2	4870	.02	97	141
18.0	.48	1.2	4684	.02	94	125
16.5	.47	1.2	4180	.02	84	99
19.5	.35	1.2	3742	.02	75	86
23.3	.27	1.2	3409	.02	68	79
25.5	.19	1.2	2648	.02	53	64

Sources: Production figures for beer and soft drink cans were provided by the Can Manufacturers Institute, and data on the proportion with soldered seams are from the same source. As virtually all soldered beverage cans are 2 11/16 inches in diameter by 4 13/16 inches in height, and use a full fillet, the amount of solder consumed per 1,000 soldered cans is found from data provided by Continental Can (see figure 3-5). Tin content of solder is deduced as explained in the text.

The conventional 2/98 and Tricom-type solders comprise a low-tin solder group. As indicated in table 3-4, the average composition of low-tin aerosol solders has been 1.9 percent tin since the mid-1950s.

However, many products packaged in aerosol cans are incompatible with high-lead solder compounds. Contact results in product contamination, container corrosion, or both. The use of presolder or postsolder stripping, which entails coating the interior seam with phenolic, vinyl, or epoxy enamels, is the preferred, least-cost means for dealing with such incompatibilities. In many cases, though, these interior linings cannot be used because the reactivity of the product also eats away the linings. When a three-piece soldered can is used for such aerosol products, a virtually lead-free solder must be employed.

Solders with 95 to 100 percent tin, the balance antimony and several minor metallic constituents, were thus utilized for about 15 percent of all soldered aerosols from the late 1950s through the late 1960s. This figure includes the high-strength, high-tin solders that contain up to .6 percent silver. As illustrated in table 3-4, the tin consumed for this smaller population of high-tin solder aerosol cans far outweighs the tin usage of the more common low-tin aerosol containers.

In the late 1960s, the problem of solder-product compatibility underwent considerable scrutiny. Hair and antiperspirant spray mists were reportedly found to contain 20 to 150 ppm of lead which could be inhaled, particularly if the mist was light enough to linger in the air. In a letter to manufacturers, the FDA advised that the industry consider options to reduce this potential hazard. Renewed research

Table 3-4. Consumption of Solder and Tin in Solder for Aerosol Cans, 1950–1978

			Low-Tin Solder				High-Tin Solder				
Year	Total Aerosol Can Production (millions)	Proportion with Soldered Seams	Cans Produced (millions)	Solder Consumed per 1,000 Soldered Cans (pounds)	Tin Content of Solder (pounds per pound solder)	Tin Consumed (tons)	Cans Produced (millions)	Solder Consumed per 1,000 Soldered Cans (pounds)	Tin Content of Solder (pounds per pound solder)	Tin Consumed (tons)	Total Tin Consumed (tons)
1950	30	.87	26	1.2	.020	*	—	.98	.95	*	*
1951	42	.90	37	1.2	.020	*	1	.98	.95	*	1
1952	97	.91	86	1.2	.020	1	2	.98	.95	1	2
1953	140	.89	118	1.2	.020	1	6	.98	.95	3	4
1954	185	.90	155	1.2	.020	2	12	.98	.95	5	7
1955	230	.90	186	1.2	.019	2	21	.98	.95	9	11
1956	304	.92	249	1.2	.019	3	31	.98	.95	13	16
1957	340	.92	275	1.2	.019	3	38	.98	.95	16	19
1958	341	.91	267	1.2	.019	3	43	.98	.95	19	22
1959	600	.92	469	1.2	.019	5	83	.98	.95	36	41
1960	743	.91	575	1.2	.019	6	101	.98	.95	43	49
1961	786	.90	601	1.2	.019	6	106	.98	.95	45	51
1962	1042	.90	797	1.2	.019	8	141	.98	.95	60	68
1963	1001	.89	757	1.2	.019	8	134	.98	.95	56	64
1964	1220	.90	933	1.2	.019	10	165	.98	.95	71	81
1965	1577	.89	1193	1.2	.019	12	211	.98	.95	88	100
1966	1649	.88	1233	1.2	.019	13	218	.98	.95	93	106
1967	1900	.88	1421	1.2	.019	15	251	.98	.95	107	122
1968	2066	.88	1545	1.2	.019	16	272	.98	.95	115	131
1969	2305	.88	1724	1.2	.019	18	304	.98	.95	130	148
1970	2456	.88	1729	1.2	.019	18	432	.98	.95	185	203
1971	2391	.89	1660	1.2	.019	17	468	.98	.95	199	216
1972	2647	.86	1707	1.2	.019	18	569	.98	.95	242	260
1973	2722	.87	1776	1.2	.019	19	592	.98	.95	253	272
1974	2553	.86	1647	1.2	.019	21	549	.98	.95	232	253
1975	2209	.78	1344	1.2	.019	14	379	.98	.95	161	175
1976	2153	.75	1324	1.2	.019	14	291	.98	.95	123	137
1977	2016	.64	1097	1.2	.019	11	193	.98	.95	83	94
1978	2078	.52	940	1.2	.019	10	140	.98	.95	61	71

Sources: Data on aerosol can production are compiled from Chemical Specialties Manufacturing Association, and figures published in various trade journals. Figures for solder consumed per thousand aerosol cans with low-tin solder are based on figure 3-5 and the estimation that the average height of these cans remained at 4¹³⁄₁₆ inches over the period examined. Data on the proportion of aerosol cans with soldered seams, the solder consumed per thousand aerosol cans with high-tin solder, and the tin content of low-tin and high-tin solder are estimates based on information available in the trade and technical literature and on interviews with industry personnel. Tin consumption is the product of the number of cans produced, the solder consumed per thousand solder cans, and the tin content for cans produced for low-tin and high-tin solder.

*Consumption is less than one-half ton.

effort into solder stripping enamels and leaching inhibitors presented no immediate solution, and as nonsoldered aerosol capacity was then inadequate, the only recourse was to adopt the costly high-tin solders. High-tin solder usage increased from its earlier level of 15 percent to about 25 percent of all soldered aerosols during 1972–75, then declined below 15 percent by 1978 as new, nonsoldered options entered the aerosol container market.

Nonsoldered aerosol containers have always been available. Indeed, the first commercial aerosol produced was Crown Cork and Seal's nominal 12-ounce drawn tinplate container, the Spra-tainer, introduced in 1946. Though the Spra-tainer was also the first aerosol in the 6-ounce size, it rapidly ceded the market to the three-piece soldered tinplate containers introduced within months by Continental Can and American Can. The high capital costs of the Spra-tainer's drawing equipment resulted in per unit manufacturing costs some 50 percent higher than similar-sized soldered containers. The Spra-tainer retained a market niche, though, in high-pressure applications, and in luxury personal products where the unaesthetic soldered container is at a marketing disadvantage. Since 1950, the Spra-tainer has never had more than 10 percent of the aerosol container market and as of 1978 served about 2 percent of it.

Other seamless aerosol containers have been on the market: American Can's drawn tinplate Pressuremaster from 1964 to 1971, Apache Can's D&I aluminum and tinplate container since 1970, and the Peerless aluminum container since the late 1950s. Combined, these seamless containers have captured less than 10 percent of the total market for aerosol containers. The principal challenge to the soldered metal aerosols, however, has come from welded three-piece containers rather than from drawn seamless aluminum or tinplate cans. While all three-piece aerosol containers produced in the United States were soldered prior to 1970, only about 55 percent were by 1979.

Continental Can's three-piece aerosol offering is a version of its TFS can initially commercialized in 1966 for the beverage can market. It is believed that Continental Can has not expanded Conoweld aerosol container production beyond the two lines brought on stream in 1972–75, and that annual production has not exceeded 180 million units.

The diffusion of the Soudronically welded tinplate container is primarily responsible for the decline in solder usage for aerosol containers. In the first few years after Southern Can offered a welded tinplate aerosol in the early 1970s, potential clients showed only modest enthusiasm. According to a Southern Can executive, marketers of aerosol products were satisfied with the performance of the soldered tinplate container. They were reluctant to switch to a new type of can that did not offer a dramatic cost advantage and was supplied by a small, relatively unknown firm. Other can manufacturers, notably American Can and Crown Cork, however, have since adopted the Soudronic technology and are producing welded tinplate aerosol containers. In 1978, about 500 million Soudronically welded tinplate containers were fabricated.

While the material cost savings accompanying the elimination of the side seam interlock have favored the adoption of the Soudronic process, as well as that of the other nonsoldered alternatives, the substitution of nonsoldered for soldered containers has come about more for reasons of container integrity. The side seam is the weakest part of soldered containers, and has the potential to rupture violently when its pressurized contents are subjected to extreme pressure or temperature.

The Consumer Product Safety Commission has investigated this potential hazard. It favors the adoption of stronger containers that utilize excess pressure relief mechanisms. These mechanisms allow the contents to vent nonviolently through prestressed perforations in the can dome. While it is in widespread use for welded and two-piece aerosols, this safety valve has proved unreliable with the weaker soldered containers.

The shift toward stronger, nonsoldered containers has also been influenced in the latter half of the 1970s by previously unanticipated concerns. In 1974, atmospheric scientists reported the likelihood that the most common aerosol propellants, chlorofluorocarbons (CFC) 11 and 12, were depleting the earth's protective ozone layer. Following the recommendation of the National Academy of Sciences, federal authorities proposed in 1976 a ban on the use of CFC 11 and 12 that went into effect in April 1979 (though exceptions do remain). When the ban was proposed, about one-half of

the 3 billion domestically produced aerosols utilized fluorocarbons. Compressed gases, CO_2, NO_2, and N_2 were usually substituted for the banned propellants. These gases are cheaper but have considerably less expansive power than the liquid fluorocarbons. They also have vapor pressures an order of magnitude greater than the liquid fluorocarbon propellants, a condition that substantially increases the likelihood of soldered side seam failure. Consequently, aerosol fillers have turned to the stronger welded or seamless containers when these compressed gases have been substituted for fluorocarbon propellants.

Experts on aerosol technology agree that the soldered aerosol container has no future. Sometime between 1982 and 1985 the last soldered aerosol line is expected to close down.

Evaporated Milk Cans

The container utilized by the packers of evaporated milk, the solder-sealed vent hole can, has remained substantially unaltered throughout the past century. The persistence of this container would be but an interesting aside were it not that its fabrication requires about one-quarter of all the tin presently utilized in canmakers' solder.

As has been the practice since the inception of the industry, the milk producers (now principally Carnation, Pet, and Beatrice Foods) fabricate their own cans at the site of filling to minimize transport costs and maintain strict internal quality control. The container bodies are joined with the common interlocking side seam, but instead of sealing the ends with organic cements as is the case with the sanitary can, closures are clinched on over the body opening and then hermetically sealed with solder. The soldering of the ends is done by either rolling the inclined cans through a 5-foot long trough containing molten solder or by applying sections of .125-inch diameter wire solder to the preheated seam circumference. One of the ends has a small hole in the center through which the milk is injected, and which is subsequently sealed by a solder drop.

When questioned as to why the industry continues to utilize a container that all other canners abandoned in the first quarter of this century, industry representatives maintain that the all-soldered vent-hole can offers superior can integrity and quality control. They state that the sanitary can is subject to "breathing," that is, microleakage about the closures in the retorting operation, which may result in spoilage of sensitive products.

They also point out that the industry "grew up" with the all-soldered can. The capital equipment of the integrated evaporated milk canning industry is geared to its usage, and the capital investment required for a switchover to any other type of container is deemed prohibitive. While not explicitly expressed by the industry representatives, it is, no doubt, difficult to justify any major capital investment for an industry that has experienced such a prolonged secular decline. Since the peak production year of 1945, when about 4.3 billion cans of evaporated milk were produced, the industry's annual output has fallen steadily, reaching an estimated 832 million cans in 1978, as shown in table 3-5. The decline of the industry is attributed in part to the availability of fresh milk and refrigerated storage, the slumping home output of baked goods, and the growth in premixed formula for infant feeding.

Table 3-5 also indicates the other apparent determinants and the amount of solder and tin in solder consumed annually in evaporated milk cans for 1940, 1945, and 1950 through 1978. The intensity of tin use in solder per million evaporated cans shown in figure 3-7 has also fallen over time. This decline in intensity of use means that the persistent and marked decline in the amount of tin consumed

in solder reported in table 3-5 is not due entirely to the substantial drop in cans of evaporated and condensed milk produced over time.

The fragmentary evidence available leads to the conclusion that there has been no major change in the amount of solder required to fabricate the evaporated milk can since World War II. For purposes of calculating annual consumption, it is estimated that 2.5 pounds of solder are required per thousand soldered (or as they are sometimes called, the floated-end) evaporated milk cans.[9]

[9]This figure of 2.5 pounds can be derived from U.S. War Metallurgy Committee tin-in-solder estimates. The committee deduced that in 1942, evaporated milk cans required 1,900 tons of tin for solder (U.S. War Metallurgy Committee, 1943, p. 420). Assuming 40-45 percent tin solder was utilized, this tin input would make about 10 million pounds of solder, which when divided by production of 4 billion containers, yields the estimated solder consumption rate. An executive with Pet, which has used

Table 3-5. Consumption of Solder and Tin in Solder for Evaporated Milk Cans, 1940, 1945, 1950–78

Year	Evaporated Milk Can Production (billions)	Proportion of Cans with Soldered Seams	Solder Consumed per 1,000 Soldered Cans (pounds)	Solder Consumed (tons)	Tin Content of Solder (pounds per pound solder)	Tin in Solder Consumed (tons)
1940	2.78	1.00	2.5	3152	.40	1261
1945	4.32	1.00	2.5	4899	.40	1960
1950	3.26	1.00	2.5	3697	.35	1294
1951	3.17	1.00	2.5	3350	.35	1173
1952	3.13	1.00	2.5	3550	.35	1242
1953	2.88	1.00	2.5	3266	.35	1143
1954	2.83	1.00	2.5	3210	.35	1124
1955	2.88	1.00	2.5	3266	.35	1143
1956	2.83	1.00	2.5	3210	.35	1124
1957	2.78	1.00	2.5	3152	.35	1103
1958	2.59	1.00	2.5	2937	.35	1028
1959	2.59	1.00	2.5	2937	.35	1028
1960	2.44	1.00	2.5	2767	.30	830
1961	2.37	1.00	2.5	2688	.30	806
1962	2.20	1.00	2.5	2495	.30	748
1963	2.08	1.00	2.5	2359	.30	708
1964	2.11	1.00	2.5	2393	.30	718
1965	1.93	1.00	2.5	2189	.30	657
1966	1.89	1.00	2.5	2144	.30	643
1967	1.67	1.00	2.5	1894	.30	568
1968	1.43	1.00	2.5	1622	.30	487
1969	1.38	1.00	2.5	1565	.30	469
1970	1.33	1.00	2.5	1509	.29	438
1971	1.28	1.00	2.5	1452	.28	406
1972	1.23	1.00	2.5	1395	.27	377
1973	1.15	1.00	2.5	1305	.26	339
1974	1.07	1.00	2.5	1214	.25	303
1975	1.03	1.00	2.5	1168	.24	280
1976	.98	1.00	2.5	1111	.23	256
1977	.90	1.00	2.5	1021	.22	225
1978	.83	1.00	2.5	942	.21	198

Sources: Figures for evaporated milk can production are compiled from data provided by the Can Manufacturers Institute and the Evaporated Milk Association. Data on the proportion of cans with soldered seams, the solder consumed per thousand soldered cans, and the tin content of solder are estimates based on information available from the trade and technical literature and on interviews with industry personnel.

This leaves the tin content of solder as the only apparent determinant, other than the number of evaporated milk cans, to change in the postwar period. While the major manufacturers of sanitary cans adopted low-tin solders during World War II, producers of evaporated milk cans continued, at first, to use traditional 40 percent tin solder. The economic incentive to reduce tin content, presumably reinforced by the reimposition of tin allocations during the Korean War, led to an incremental reevaluation of traditional usage. Industry representatives relate that 35 percent tin solder prevailed during the 1950s, dropping to 30 percent in the 1960s. During the 1970s, tin content was reduced incrementally, with current tin content averaging 20 percent. The industry representatives clearly identified the cost of tin as the motivating factor for this change in solder consumption.

Comparative Effects of Apparent Determinants

The comparative effects of changes in the apparent determinants of tin consumption in can solder are illustrated in figure 3-8. Shaded histograms are shown for actual tin usage in 1941 and 1950 for fruit and vegetable cans and evaporated milk cans, and in 1950 for beverage cans and aerosol cans. Prewar histograms for beverage or aerosol cans are not presented since the output of these can types was insignificant before the war.

To the right of each histogram indicating actual usage in 1941 or 1950 is a histogram that shows how much tin would have been consumed in the subsequent period—either 1950 or 1978—had only those apparent determinants acting to increase tin use changed. This second histogram also illustrates (with stippling) the multiplicative effects of change by two or more apparent determinants. (The multiplicative effect has been more thoroughly defined in the preceding chapter.) To the right of each histogram indicating these additive impacts is another which nets out the individual and multiplicative effects of those apparent determinants tending to reduce tin consumption. The remaining portion of this histogram, which is shaded, indicates the actual level of tin consumption in either 1950 or 1978.

In a number of instances, no multiplicative effect exists since only one apparent determinant tends to increase or decrease tin consumption. In the case of evaporated milk cans, there is no apparent determinant tending to increase tin consumption between 1950 and 1978. As a result, there is only one histogram for 1978 which shows the impact of those apparent determinants which acted to decrease tin consumption.

Looking at the magnitude of the changes indicated in figure 3-8, one finds that the 1941–50 reduction in the tin content of fruit and vegetable can solder has had the largest impact of all on tin in solder usage. Low-tin solders were first evaluated in anticipation of the strategic vulnerability of tin supply, and then adopted when tin supply was indeed problematic. However, they diffused rapidly in large part because they offered equivalent to superior service at greatly reduced material cost compared with the prewar 40-50 percent tin solders.

Material cost has been a major motive behind the incremental reduction in the tin content of evaporated milk can solder, which occurred largely after World War II. However, of even greater impact than changing composition on the total level of tin in solder usage has been the precipitous decline in the output of evaporated milk.

With beverage containers, the important determinants have been the growth in can production, and the declining proportion of cans with soldered side seams. The cost of solder constitutes only a minor fraction of the total material cost savings effected by adopting nonsoldered containers. As a consequence, the cost of tin in solder has not been a factor in the switch away from soldered cans. The enhanced integrity and versatility of nonsoldered beverage containers, along with lower overall costs, have been of much greater importance.

The use of nearly pure tin solders for portions of the aerosol and fruit and vegetable can markets has been the only instance where the level of tin use in can solder has increased significantly in the postwar period. In both cases, either the contents or container integrity were imperiled by high-lead solders. High-tin solders, though costly, represented the most readily available remedy to the problem.

With aerosols, nonsoldered alternatives have recently penetrated the market, reducing but not yet eliminating the need for high-tin solders. These alternatives, particularly the welded tinplate container, have also captured a large share of those aerosol products previously packaged in low-tin soldered containers due to their superior can integrity at comparable cost. A similar course of events may soon follow in the fruit and vegetable container market.

Motor Vehicle Radiators

Only a fraction of the energy generated by the combustion of fuel in the engine of a motor vehicle is translated into motive power. The remainder is converted into waste heat which must be removed from the engine; otherwise, metal warpage may seriously damage moving parts. Within a few years of the automobile's inception, water-cooling designs replaced direct air-cooling of engines and have remained in near universal use ever since.

---

a wire-soldering technique at least since World War II, provided a current estimate of net solder consumption that falls in the range of 2.25-3.33 pounds per thousand. This wide range is reportedly due to the variability of the surface conditions of tinplate stock; lower quality tinplate requires more solder to achieve a reliable, hermetic seal.

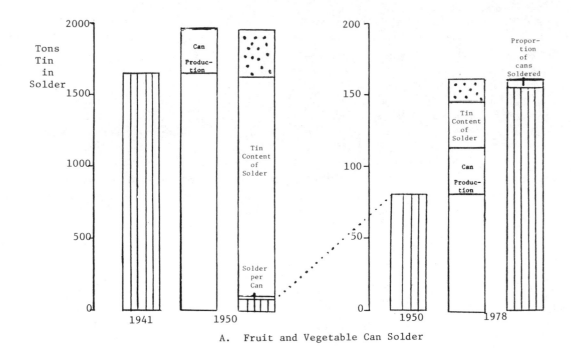

A. Fruit and Vegetable Can Solder

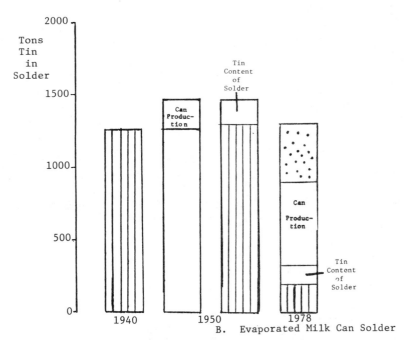

B. Evaporated Milk Can Solder

Figure 3–8. Effects of changes in the apparent determinants on tin consumption in can solder.

Engine parts susceptible to waste heat damage, the cylinders and valve parts, are jacketed with passageways. A water-based solution circulates through the jacketing and absorbs the waste heat and removes it from the engine. The heated fluid then passes through the radiator where the heat is transferred primarily by conduction and convection to a cooler medium, the onrushing airflow. The cooled water is then recirculated back to the engine jacketing.

Copper and brass have been used to construct engine radiators since the earliest days of the automobile industry. Copper and brass are highly thermal conductive, easy to fabricate, and are readily joined with tin-lead solders. Solder acts to join radiator components (waterways, airways, tanks, and supporting frame) in structural and thermal contact.

To reduce material and production costs and to forestall penetration of the aluminum radiator, which has been under

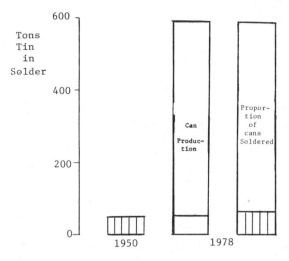

C. Beverage Can Solder

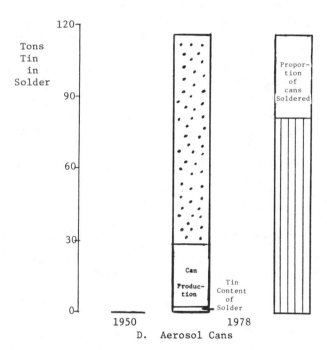

D. Aerosol Cans

Figure 3–8. *continued*

development since World War II, new designs and performance requirements have evolved over the past four decades for the copper-brass radiator, involving the use of new soldering techniques, different per-unit usage rates, and changes in solder composition.

Radiator Design and Fabrication

Various radiator designs have been employed to dissipate automobile engine heat. In the postwar period, the most common type has been the tube and corrugated fin, or pack construction radiator. Figure 3-9 illustrates the design of

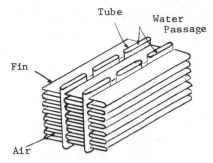

Figure 3–9. **Tube and corrugated fin radiator core.**

this radiator's core, in which the bulk of the heat transfer from the water to the airstream takes place.

The flat oval tubes carry the heated water through the core. A radiator may have from 35 to 75 tubes, depending upon heat removal requirements. The tubes are formed from cartridge brass strip 1¼-1½ inches wide and .005 inch thick. As indicated in figure 3-10, the strip is flanged at the ends, then bent at the center to form an envelope, then the overlapped flanged ends are tightly crimped into a lockseam. Either before or after the tube is sectioned into lengths of 20-26 inches, the tube exterior is fluxed, solder-coated in a molten solder bath, and then wiped of excess solder. Forbes (1965, p. 12) states that this coating is between 0.6-1.2 thousandths of an inch thick.

Fins are now made from tough pitch anneal-resistant electrolytic copper. In the tube and corrugated fin construction, the fin stock sheet is corrugated into a tight isoclinal pattern, then inserted between the regular-spaced columns of tubes with the fold of the fins in line with the airflow. The hinges of the folds are juxtaposed to the tubes, forming the tube-to-fin joint, which are joined by reflow (or sweating) of the tube solder coating upon heating of the core. Thus, the tube solder coating has two functions: it seals the tube lockseam, and sweat solders the tubes of fins in structural and thermal contact.

The matrix of tubes and fins is known as the radiator core. To reinforce the core, a perforated plate .025-.030 inch thick is placed over the protruding tubes at both ends of the core. This plate (or header) is attached by infilling the clearance about each tube end with solder. Numerous techniques have been employed to seal this set of joints;

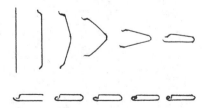

Figure 3–10. **The radiator tube-forming sequence.**

pouring a carefully measured amount of solder over the plate or injecting a solder stream along the face of the header plate are currently employed.

Brass inlet and outlet tanks are then attached to the top and bottom header plate with solder. As with the header, the soldering technique has evolved from an all-hand operation to a highly mechanized, semiautomated one. To prevent the remelting of previously soldered joints, particularly the tube-to-header joint-set, most manufacturers consider it necessary to use solders with lower melting temperatures in attaching the tanks to the core.

Figure 3-11 diagrams the primary components of a radiator—core, header plates and tanks—and indicates the site of tin-lead solder application. Other tasks in the fabrication of a tube and corrugated fin radiator require the use of tin-lead solders: the attachment of a tank inlet and outlet pipe, draincock, and side support brackets, and solder for in-shop repair. These account for substantially less solder usage than the previously identified tasks, and are not considered in this analysis.

The antecedent to the tube and corrugated fin (or pack construction) radiator was the tube and flat-fin core, shown in figure 3-12. As its name implies, the fins are thin, flat copper foils that are perforated with a pattern of holes, shaped and sized to fit around the tubes. The solder-coated tubes are inserted through the holes, and the fins are stacked up, 7 to 14 fins per inch. As with the pack construction, tube-to-fin bonding is done by reheating the solder coating on the tubes. While very popular with U.S. automakers immediately after World War II, the tube and flat-fin radiator was rapidly displaced by the less labor-intensive pack construction.

Up through the 1940s and into the early 1950s, the cellular or interleaved film radiator, illustrated in figure 3-13, was also widely used. Each vertical water passage is made from two stamped copper sheets which have furled edges that make a narrow envelope when brought together. The metal envelope is corrugated to give maximum surface area given

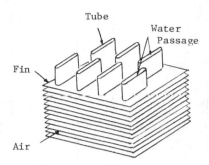

Figure 3–12.   Tube and flat fin radiator core.

the vertical height. Corrugated copper fins are then juxtaposed against the waterway envelope for maximum contact. A sandwich of corrugated fins and waterways is clamped into a jig and is then ready for soldering. Soldering is achieved by dipping the front and back face of the core into a molten solder bath 1/8 to 1/2 inch deep. The solder is drawn up into the seam of continuous contact between the metal waterway and airways by capillary action. No header plate is necessary with the cellular core, and both tanks are often clamped onto the core prior to face-dipping and thus soldered to the core in the same operation that joins the core components.

While strong and flexible, the cellular radiator was liable to failure when subjected to intermittent internal pressures. For this reason, it was phased out in the late 1940s and early 1950s when the pressurization of radiators was adopted, though the cellular heat exchanger is still used today for automobile heater cores.[10]

Solder Usage Before and During World War II

In the prewar automobile, the dip-soldered cellular radiator was far more common than the sweat-soldered tube and fin type. During the decade preceding America's entry into World War II, it appears that some manufacturers reduced

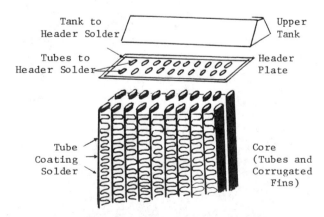

Figure 3–11.   Upper half of a tube and corrugated fin radiator with sites of solder usage indicated.

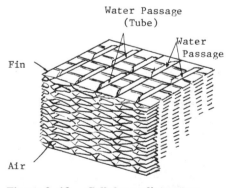

Figure 3–13.   Cellular radiator core.

[10]Cellular edge-dipped heater cores represent a large market for solder. Each heater core requires up to 2 pounds of solder. Industry-wide, tin content ranges from 3 to 20 percent. Thus, approximately 1,200 tons of tin were likely used for heater cores in 1978.

the tin content of core dipping solder from 40-50 percent tin (Hiers, 1931, p. 250) to as low as 25-30 percent (Houssner and Johnson, 1946, p. 250). For all other tasks in the manufacture of both cellular and tubular radiators, the all-purpose 40-50 percent tin solder continued to be used.

Considerable differences exist in the published estimates of the amount of solder and tin in solder used in the prewar radiators. The National Academy of Sciences (1941) Advisory Committee on Metals and Minerals estimated that dip soldering then required about 2 pounds of tin per radiator. Given that the lowest tin content of dipping soldering then in use was 25 percent, at most 8 pounds of solder were used per radiator. However, according to Houssner and Johnson (1946, p. 249), who were both with Chrysler, prewar automotive radiator manufacture required only 1.2 pounds of tin per unit. This implies that, at most, 4.8 pounds of solder could have been used per cellular radiator, with actual solder usage presumably less as higher tin content solders were required for attachment of the tanks and fittings.

For the purpose of calculating prewar material consumption in radiator fabrication, the more precise estimates of Houssner and Johnson were favored. Specifically, it was surmised that an average automobile radiator contained 1.5 pounds of tin and 4.25 pounds of solder, implying an average tin content of about 35 percent.

With the pending entry of the United States into the world war and mounting concern over tin availability, the National Academy of Sciences Advisory Committee on Metals and Minerals suggested a measure to limit tin consumption in radiator fabrication. Noting that glycol-cooled aircraft engine radiators utilized dipping solder composed of 94-95 percent lead and 5-6 percent silver to make joints of superior strength, the committee recommended that automotive radiators adopt similar solders. To moderate silver usage, it suggested that tin-free solders with somewhat lower silver content could be utilized.[11]

Indeed, silver-lead solders were adopted for motor vehicle radiator dip-soldering, but apparently only by the Canadian Army for truck radiators (Houssner and Johnson, 1946, p. 249). The U.S. manufacturers of wartime motor vehicle radiators instead chose to comply with restrictions on tin availability by incrementally reducing the tin content of their solders. Manufacturers of tube and fin radiators reduced tin content from the traditional 40-50 percent to the amount permitted by the War Production Board, 32 percent. For the dip-soldering of cellular radiators, 15 percent tin was found to suffice. The attachment of the tanks to the core was performed by semiautomated procedures requiring 30 percent tin solder, but hand-soldering operations, such as the attachment of fittings, continued to require 40 percent tin solder.

Houssner and Johnson estimate that as a result of these changes, the amount of tin in both tubular and cellular radiators was reduced to about .75 pound. While the authors do not so state, this implies that, at least in the case of the tubular radiator, the amount of solder used also fell. With the decline in average tin content of solder to 32 percent, noted above, the .75 pound of tin used suggests that each radiator contained 2.3 pound of solder. Unfortunately, there is no other evidence available to confirm this conclusion.

Wartime exigencies also had an impact, albeit a long-term, indirect one, on the proportion of soldered to non-soldered automobile radiators. To save every ounce of weight possible, many U.S. military aircraft were equipped with brazed aluminum heat exchanges (intercoolers), made by Trane, instead of heavier copper ones. Immediately after the war, the Office of Technical Services obtained machinery used in Germany during the war to make welded aluminum aircraft intercoolers, as well as two machines producing aluminum motor vehicle radiators (*Steel*, 1946, p. 102). These developments, as well as their own wartime research, led engineers at Harrison Radiator Division of General Motors to undertake a program for the mass production of an aluminum automobile radiator.

Radiator Solder Usage Since 1950

Table 3-6 indicates the estimated values of the apparent determinants for 1941 and 1950 through 1978: the number of new automobile radiators (X_t), the proportion of soldered to nonsoldered radiators (c_t), the consumption of solder per soldered radiator (b_t), and usage of tin per pound of solder (a_t). It also shows their product (Q_t), the amount of tin used in new automobile radiator solder. This same exercise is repeated in table 3-7 for new light truck radiators, where a light truck is defined as a truck with gross vehicle weight of 10,000 pounds or less. Two-thirds (in the early 1950s) to nine-tenths (in the late 1970s) of all U.S.-produced trucks have been in this 0-10,000 pound weight class.[12]

From table 3-6 it is calculated that while solder usage for passenger vehicle radiators has increased nearly .5 percent annually since the early 1950s, tin in solder usage has declined at an average 2.1 percent annual rate. With light trucks, radiator solder usage has increased at a 5.7 percent average annual rate, while tin in solder usage has increased only 2.3 percent. As the production of new passenger automobile and light truck radiators (exogenous variable X_t) has increased at an average annual rate of 1.6 and 6.8 percent annually, the other apparent determinants of usage have counteracted this growth trend and resulted in the comparatively lower growth of radiator solder and tin in solder usage. Figure 3-14 illustrates that the tin use in solder per

---

[11]Note that this same solution, the adoption of silver-lead solders, was also recommended for tin savings in canmakers' solder.

[12]Trucks in this vehicle class include compact and conventional pickups, vans, passenger carriers, utility vehicles, and station wagons on truck chassis.

Table 3-6. Consumption of Solder and Tin in Solder for New Passenger Automobile Radiators, 1941, 1950–78

Year	New Passenger Vehicle Radiator Production (thousands)	Proportion Soldered	Solder Consumption		Tin in Solder Consumption	
			Pounds per Soldered Radiator	Total (tons)	Pounds per Pound Solder	Total (tons)
1941	3780	1.00	4.25	7287	.35	2550
1950	6666	1.00	3.20	9676	.27	2613
1951	5338	1.00	3.03	7324	.29	2124
1952	4321	1.00	2.85	5586	.31	1732
1953	6167	1.00	2.67	7483	.33	2469
1954	5559	1.00	2.59	6525	.32	2088
1955	7920	1.00	2.35	8197	.33	2705
1956	5816	1.00	2.35	6199	.30	1860
1957	6113	1.00	2.35	6538	.29	1896
1958	4258	1.00	2.35	4538	.28	1271
1959	5591	.99	2.35	5961	.25	1490
1960	6675	.99	2.35	7114	.23	1636
1961	5443	.99	2.35	5800	.22	1276
1962	6933	.99	2.35	7389	.22	1626
1963	7638	.99	2.35	8144	.22	1792
1964	7752	.99	2.35	8264	.20	1653
1965	9306	.99	2.35	9920	.19	1885
1966	8598	.99	2.35	9165	.19	1741
1967	7437	.99	2.35	7927	.18	1427
1968	8822	.99	2.35	9404	.18	1693
1969	8224	.99	2.35	8766	.18	1578
1970	6547	.99	2.35	6979	.18	1256
1971	8585	.99	2.35	9142	.18	1646
1972	8824	1.00	2.35	9406	.18	1693
1973	9658	1.00	2.35	10295	.18	1853
1974	7331	1.00	2.30	7648	.18	1377
1975	6713	1.00	2.25	6856	.16	1096
1976	8498	1.00	2.20	8480	.16	1357
1977	9201	.99[a]	2.20	9137	.16	1462
1978	9165	.98[a]	2.10	8599	.16	1376

Sources: Data on U.S. production of new passenger motor vehicles are from the Motor Vehicle Manufacturers Association, *Facts and Figures* (annual). (One new passenger vehicle produced requires one new radiator.) Other data are estimated, as described in the text.

[a]In 1977 production of the VW Rabbit was initiated in Pennsylvania. While the Rabbit is produced in the U.S., its nonsoldered aluminum radiator is imported. Nonetheless, production of the Rabbit has been factored into estimates of the proportion of radiators soldered.

motor vehicle radiator has been persistently declining since the early 1950s.

TIN CONTENT OF RADIATOR SOLDER.   To a large extent, the noted trend in intensity of tin use for motor vehicle radiators can be explained by changes in the composition of radiator solder. These changes have occurred in response to the redesign and respecification of the operating parameters of the automobile engine radiator.

The physics of heat transfer establish that the heat rejection capability of the radiator can be improved by increasing the temperature difference between the fluid in the radiator and the air passing about it. To prevent the fluid from boiling, the heat exchanger system can be pressurized.

Pressurization of the automobile radiator concurrent with an increase in operating temperatures was initiated in the 1940s, and was rapidly adopted during and immediately after the Korean War. As mentioned previously, the cellular radiator lacked the structural integrity to withstand cyclic

stress of intermittent pressurization and was thus withdrawn from use as an engine radiator during this period.

For passenger vehicles, the postwar dip-soldered cellular radiator, utilizing some 4¼ pounds of 15-20 percent tin solder, was thus supplanted by the sweat-soldered tubular radiator, which in the late 1940s and early 1950s required 2⅓ pounds of an average 35 percent tin solder. Trucks of less than 6,000 pounds used an estimated 5 pounds per dip-soldered and 2.5 pounds per sweat-soldered radiator, whereas trucks of 6,000 to 10,000 pounds are estimated to have required 7 pounds of solder per dip-soldered, and 3.5 pounds of solder per sweat-soldered unit. Light trucks and passenger vehicles are considered to have followed the same trends regarding solder composition. Solder consumption per radiator was thus reduced by half, but the average tin content of solder increased, as shown in table 3-6, for the period 1950–55. As a result, tin consumption remained the same, or even increased slightly. The average sweat-solder com-

Table 3-7. Consumption of Solder and Tin in Solder for New Light Truck Radiators, 1941, 1950–78

	0–6,000 Pound Vehicle Weight Class				6,000–10,000 Pound Vehicle Weight Class				Combined	
Year	Truck Radiator Production (thousands)	Proportion Soldered	Solder Consumed per Soldered Radiator (pounds)	Solder Consumed (tons)	Truck Radiator Production (thousands)	Proportion Soldered	Solder Consumed per Soldered Radiator (pounds)	Solder Consumed (tons)	Tin Content (pounds per pound solder)	Tin Consumed (tons)
1941	455	1.0	5.00	1032	113	1.0	7.00	359	.35	487
1950	525	1.0	3.50	833	131	1.0	4.90	291	.26	292
1951	560	1.0	3.25	826	140	1.0	4.55	289	.27	301
1952	413	1.0	3.00	562	108	1.0	4.20	206	.30	230
1953	524	1.0	2.75	654	131	1.0	3.85	229	.32	282
1954	454	1.0	2.63	541	113	1.0	3.68	188	.32	233
1955	524	1.0	2.50	594	131	1.0	3.50	208	.33	265
1956	442	1.0	2.50	501	110	1.0	3.50	175	.30	203
1957	462	1.0	2.50	524	116	1.0	3.50	184	.29	205
1958	382	1.0	2.50	433	96	1.0	3.50	152	.28	164
1959	489	1.0	2.50	555	122	1.0	3.50	194	.25	187
1960	503	1.0	2.50	570	132	1.0	3.50	210	.23	179
1961	526	1.0	2.50	596	128	1.0	3.50	203	.22	176
1962	615	1.0	2.50	697	159	1.0	3.50	252	.22	209
1963	729	1.0	2.50	827	185	1.0	3.50	294	.22	246
1964	830	1.0	2.50	941	203	1.0	3.50	322	.20	253
1965	940	1.0	2.50	1066	254	1.0	3.50	403	.19	279
1966	966	1.0	2.50	1095	278	1.0	3.50	441	.19	292
1967	918	1.0	2.50	1041	276	1.0	3.50	438	.18	266
1968	1083	1.0	2.50	1228	381	1.0	3.50	605	.18	330
1969	1134	1.0	2.50	1286	418	1.0	3.50	664	.18	351
1970	1014	1.0	2.50	1150	394	1.0	3.50	626	.18	320
1971	1184	1.0	2.50	1343	488	1.0	3.50	775	.18	381
1972	1449	1.0	2.50	1643	599	1.0	3.50	951	.18	467
1973	1662	1.0	2.50	1885	758	1.0	3.50	1203	.18	556
1974	1392	1.0	2.50	1579	696	1.0	3.50	1105	.18	483
1975	999	1.0	2.50	1133	952	1.0	3.50	1511	.16	423
1976	1218	1.0	2.50	1381	1401	1.0	3.50	2224	.16	577
1977	1173	1.0	2.50	1330	1803	1.0	3.50	2662	.16	671
1978	1193	1.0	2.50	1353	2140	1.0	3.50	3397	.16	760

Sources: Data on U.S. production of new light trucks 1960–78 are from the Motor Vehicle Manufacturers Association of the United States Inc. (1979) *Information Handbook*, p. 8. Data for 1941, 1950–59 are estimated from information given in this same source (p. 22). Other data are estimated, as described in the text.

position of 35 percent was comprised of roughly equal proportions of 25 percent tin solder for tube coating, 40 percent tin solder for the tube-to-header joint set, and 40 percent tin solder for other tank to core joint seam (inference from Ricker, 1948, p. 81).

During the Korean War, the trend toward pressurization was accelerated by the necessity to reduce the amount of copper used in radiators due to tight government controls over the red metal. Copper gauge and surface area were both reduced, while operating temperature and pressure were increased in order to maintain heat rejection per unit time through a smaller heat exchanger. After the Korean War, although copper was readily available, cooling system engineers continued to pare down the weight of the average radiator to minimize cost. Figure 3-15 illustrates the post-1950 trends in operating temperature, pressure, and radiator

weight, as estimated on the basis of information from the trade press.

Under normal operating conditions, these elevated temperatures and pressures pose no risk of abrupt rupture at the radiator's soldered seams, but they do accelerate solder creep rates, leading to potential failure. Unlike most other solder bulk strength properties, creep strength increases over the range of 0-40 percent tin as tin content is lowered; thus at elevated operating temperatures, the solders with low tin content and higher melting temperature have improved creep strength properties. As noted with can seam solder, the presence of 0.5 to 1.0 percent silver further strengthens low-tin solders.

With this knowledge, radiator manufacturers introduced in the mid- to late 1950s low-tin, silver-bearing solders to replace the 30-40 percent tin solders used to join the tubes

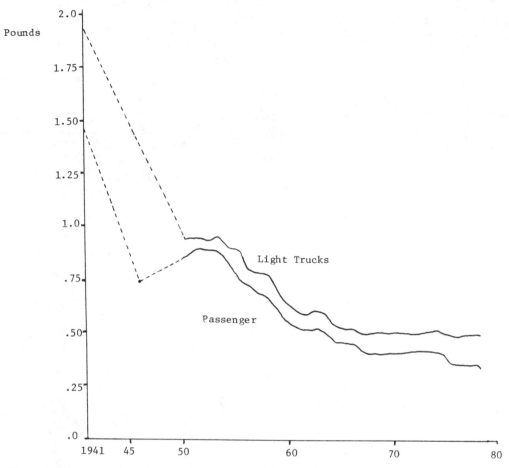

Figure 3–14. Pounds of tin in solder per new light trucks and passenger vehicle radiator, 1941–78. [Datum for 1946 passenger vehicles is from Houssner and Johnson (1946). All other data are from tables 3–6 and 3–7.]

to the header. By the mid-1960s, solders averaging 2.5 percent tin, 0.5 percent silver, and the balance lead, plus minor additives, were in widespread use for this joint set. This development reduced tin usage nearly 0.3 pound per radiator from 1950 levels, and is responsible for most of the drop in the average tin content of solder between 1955 and 1966 recorded in table 3-6.

While they were adopted for their superior performance under more rigorous operating conditions, the low-tin, silver-bearing solders also had a considerable material cost advantage over the 30-40 percent tin solders that they replaced, costing on average 50 percent less per unit weight. Although the decline in tin usage for tubes-to-header solder is temporally correlated with the rising price of tin relative to lead shown in figure 3-1, material costs were not the major factor causing this switch away from high-tin solders.

According to radiator manufacturers, material costs have been a more important influence in the evolutionary reduction in tin content of tube coating and tank-to-header solders. Changes in solder composition for these applications, though, have been constrained by the necessity to maintain tem-

perature differentials in the sequential process of soldering the various components of the radiator.

Tube coating solder, which seals the tube lock seam and joins tubes to fins, ranged from 25-30 percent tin in 1950. As solder with higher melting temperature was introduced for tube-to-header solder, it was necessary to raise the melting temperature of tube coating solder to ensure that the tubes-to-header soldering, which occurs later in the production cycle, did not cause tube leaks to develop. Apparently, though, some latitude in the soldering process exists, since the tin content of tube coating solder currently used by the Big Three automakers ranges from 5 to 23 percent.

At any rate, for the purpose of estimating material consumption, tube coating solder is assumed to have averaged 15 percent tin since 1975, with the decline from 25-30 percent having occurred gradually over the 1950–75 period. This change in the tin content of tube coating solder accounted for slightly over 20 percent of the postwar decline in the tin content of radiator solder shown in table 3-6.

Since the decline in the tin content of tube coating and tube-to-header solders, the joining of top and bottom tanks

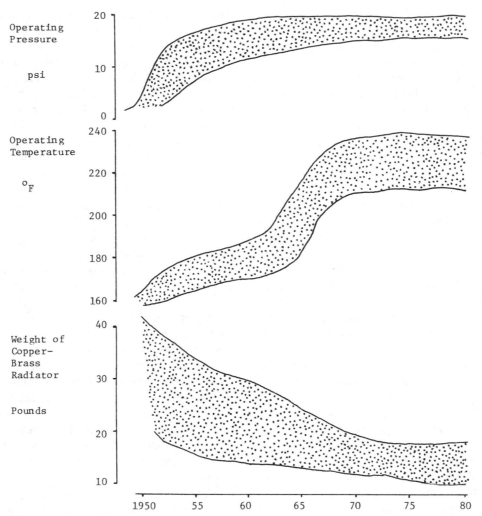

Figure 3–15. Operating pressure, temperature, and weight of the copper-brass passenger vehicle radiator, 1950–80. [From trade journal articles and interviews.]

to the header plate now requires more tin than both other tasks combined. While throughout the 1950s 40 percent tin solder was typical for tank-to-header solder, 30 percent was in common use by the late 1960s and remained the standard through 1978. The principal physical specification for the tank-to-header solder is a high degree of fluidity. It must also possess a melting temperature lower than that of the other radiator solders, lest the heat of tank soldering result in core leaks.

Early in 1979, the high price of tin caused radiator fabricators to reconsider their usage of 30 percent tin tank-to-header solder. Apparently during 1979–80, some radiators had their tanks attached with 10-20 percent tin solder, and one expert believes 5 percent tin solder is feasible. Also being seriously considered is the replacement of soldered copper tanks with mechanically attached plastic ones. Should the plastic tank or another alternative that is being tested,

the arc-welded tank, be adopted, total tin in solder usage would drop below 0.1 pound per radiator.

SOLDER USAGE PER SOLDERED RADIATOR. The obsolescence of the dip-soldered cellular radiator, discussed above, and its replacement by the sweat-soldered tube variety had a major impact on solder use per radiator, causing passenger vehicle radiator solder consumption to drop from 4.25 to 2.35 pounds. This switchover was completed by about 1955. From that time through the 1960s and early 1970s, the trend toward lighter, smaller radiators was offset by the use of higher lead content (denser) solders. As a result, the solder used per radiator remained fairly constant at about 2.35 pounds between 1955 and 1973.

From 1974 on, the trend toward smaller passenger vehicles and smaller engines has reduced solder usage. Smaller engines have lower heat rejection requirements; hence, the radiator needs fewer tubes (with 5 grams of solder coating per tube) and a narrower tank to accommodate the task. For this reason, solder usage per radiator is estimated to have declined since 1973 by about 10 percent to 2.10 pounds. The three radiator soldering tasks analyzed continue to divide the 2.10 pounds of solder about equally.

The Ford Motor Company has been investigating the replacement of soldered copper tanks with mechanically attached plastic ones. Should the plastic tank or another alternative that is being tested, the arc-welded-on tank, be commercialized, it is estimated that 1.4 pounds of solder averaging 5-10 percent tin will be used for the remaining tube coating and tube-to-header joining tasks.

With light trucks, the trend since the late 1960s has been toward a larger output share for vehicles in the 6-10,000 pound class, requiring 3.5 pounds of solder per radiator, with a concomitant decline in market share for vehicles less than 6,000 pounds, requiring 2.5 pounds of solder. The resultant increase in average solder consumption per light truck has offset the decline in the average tin content of radiator solder, hence tin consumption per light truck radiator has been static at .5 pound per unit since the late 1960s. (See figure 3-14.)

NONSOLDERED RADIATORS. Periodically since 1951, the trade press has proclaimed that the automobile radiator was on the verge of a radical transformation through the substitution of aluminum for copper-brass stock. Since aluminum is virtually unsolderable and other metal-joining techniques must be used, the decision to replace the copper-brass radiator with an aluminum one would profoundly affect tin-lead solder demand.

As with all material inputs, the demand for the metallic components of an automotive radiator is a function of the desirable physical characteristics they yield. For a heat exchanger, the principal attribute desired is thermal conductivity. So, when comparing two competing metals, thermal conductivity per unit of material cost and per unit effective (or life) cost must be considered.

It is a simple task to compare the material costs of heat exchanger materials (copper, brass, and aluminum) with their heat rejection capability. One need consider only relative material prices (per unit weight) and the ratio of thermal conductivity (per unit volume) to density (weight per unit volume). When these calculations are carried out for copper, brass, and aluminum, they indicate that, if all other considerations are equal, aluminum should be used instead of copper when the price of copper exceeds 58 percent of the price of aluminum. Aluminum should similarly be used instead of brass when the price of brass exceeds 19 percent of the price of aluminum. Since the price of copper has been greater than the price of aluminum since 1947, with the exception of 1958, and the price of brass has been par with aluminum, such calculations strongly favor the use of aluminum for automobile radiators. Compound aluminum's relative price advantage with copper's periodic bouts with uncertain supply and erratic price performance, and the case for the aluminum radiator becomes even more impressive. Still, all but a small fraction of postwar radiators have been made of copper and brass. Obviously, other factors have offset these advantages of aluminum.

It is certainly not for lack of interest, research, or development activity that the aluminum radiator is so little used. Harrison Radiator Division of General Motors and Alcoa have spent large sums on the development of the aluminum automobile radiator since the late 1940s. Other firms have since undertaken their own development projects. During the 1950s, General Motors field-tested over a thousand aluminum radiators, and in the 1960s produced on a pilot line about 250,000 aluminum radiators for the Corvette before the project was terminated. The Corvette aluminum radiator never could compete with the copper-brass radiator in terms of total production costs or performance reliability, though it did reduce the weight in front of the axle, which is an important design consideration for a high-performance sports car. With the current effort to limit motor vehicle weight, the aluminum radiator is again under intense industry study.

A major problem with the aluminum radiator is its susceptibility to corrosion pitting from contact with the coolant fluid, as well as from rainwater and salt. The use of thicker gauge aluminum stock reduces the incidence of pitting, but also diminishes the material cost advantage of aluminum. Corrosion inhibitors in solution with the coolant require more frequent replenishment than many car owners would remember.

Once an aluminum radiator develops a leak, repair is far more difficult than with the readily solderable copper-brass unit due to the aluminum alloy's refractory oxide skin. Instead of being repaired, leaking Corvette radiators were normally replaced, typically with copper units. Repair with adhesives has as yet proved unreliable, and various plating-soldering techniques require far more skilled labor than copper-brass unit repair. Advocates of the aluminum radiator maintain that reliable low-cost repair is in the offing (Morley and Wilkinson, 1978), but radiator production engineers consider the uncertainty of reliable servicing a significant factor in inhibiting commercialization.

The development of new fabricating techniques has reduced the variable costs of producing aluminum radiators significantly and costs no longer pose the barrier to adoption they once did. Previously, the Corvette aluminum radiator and the earlier trials of General Motors were all dip-brazed; that is, dipped in a molten salt flux bath that cleans off the oxide skin, allowing preplaced brazing metal to then join the components. Also, the corrosive flux had to be completely cleaned off the unit. These operations resulted in component joining costs three to four (1972) dollars higher than a comparable copper-brass radiator (National Materials Advisory Board, 1972, p. 67). Other brazing techniques

A. Passenger Vehicles

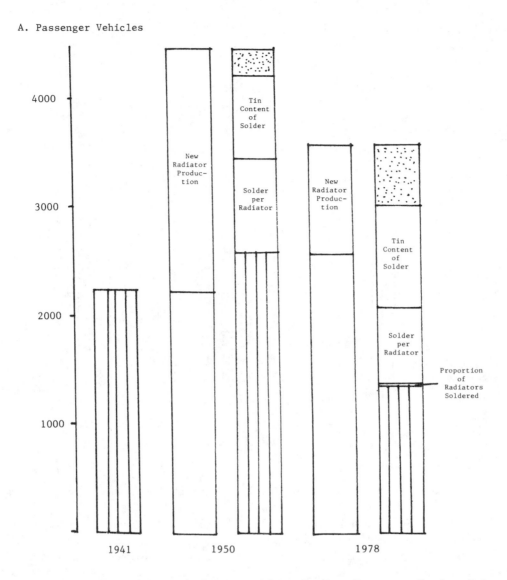

Figure 3–16. Effects of changes in the apparent determinants on tin consumption in radiator solder for new passenger vehicles and light trucks.

were investigated that utilized less flux but did not have the potential for scale-up to high-speed, production-line processing. Finally, however, the persistence of the aluminum companies' research efforts paid off because the vacuum-brazing method developed in the late 1960s requires no flux and is reported to have production costs comparable to the copper-brass radiator.

Since then, the capital cost of conversion has been the principal production cost element forestalling the adoption of the aluminum engine radiator, estimated in 1972 at 200 million dollars (National Materials Advisory Board, 1972, p. 67). Additionally, the automotive radiator industry has thousands of man-years of expertise, or invested human capital, in the fabrication of copper-brass radiators.

Of all the automobiles produced in the United States, as of 1979 only the Volkswagen Rabbit, assembled in New

Stanton, Pennsylvania, uses an aluminum radiator. Output in 1979 was about 167,000 units. Previously, VW vehicles were air-cooled, so the company had no commitment to the copper-brass radiator. The radiators are fabricated in Europe under an output-sharing scheme with the developer of the process, SOFICA, a French firm. All joints are made mechanically, without soldering or brazing, and the radiator uses a plastic tank. When damaged or defective, the unit is replaced. In the 1981 model year General Motors started a new aluminum radiator testing program and has featured a new aluminum radiator in the 1982 Camaro.

While the aluminum automotive radiator has yet to penetrate the automobile radiator market in the United States, except to the limited extent described, the threat of substitution has prompted copper producers and others with interests in preserving the copper-brass radiator to join in the

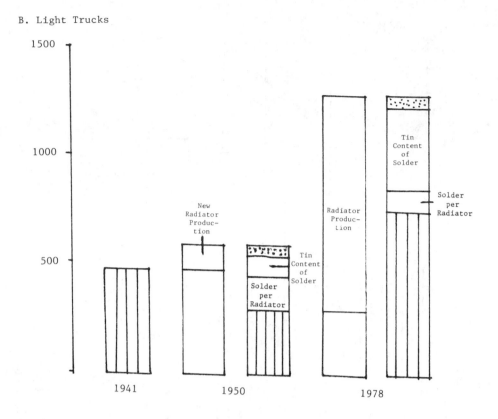

B. Light Trucks

Figure 3–16.   *continued*

funding of projects headed by the Copper Development Association (CDA) and the International Copper Research Association (INCRA) that are aimed at improving the reliability and reducing the cost of copper-brass radiators.

Of their many process inventions, several, in particular, may affect the amount of solder used per radiator and its tin content, including the butt welding of tubes, the plasma-arc welding of header tanks and filler necks, the Marston radiator which incorporates all of the above features, and zinc-based solders. Reportedly, the Marston radiator (named for the developing company which is the largest radiator manufacturer in Britain) uses 80 percent less tin-lead solder, 50 percent less copper-brass, and 35 percent less labor than comparable European practices (Copper Development Association, n.d., Data Sheet 804/7). Currently, the radiator is undergoing field tests by Chrysler of Britain and other European automakers. As for the other projects mentioned, only the welding of filler necks, which eliminates the need for a minor amount of solder, has been commercialized in the United States. Still, the competition between the soldered copper-brass radiator and the non-soldered aluminum radiator, through the stimulus it provides for new technology, may significantly affect other apparent determinants of tin consumption, even if the soldered copper-brass radiator continues to dominate the U.S. automobile market.

Comparative Effects of Apparent Determinants

Figure 3-16A illustrates the amount of tin consumed in new passenger vehicle radiators in 1941, 1950, and 1978, as well as the individual and multiplicative effects of the apparent determinants on the use of tin tending to increase or decrease tin usage between 1941 and 1950 and between 1950 and 1978. Figure 3-16B illustrates this same information for new light truck radiators.

For both the 1941–50 and the 1950–78 periods, the only apparent determinant acting to increase tin in radiator solder usage for both autos and light trucks was the number of radiators produced. Each of the other three apparent determinants has acted to reduce tin use. During the 1941–50 interval, the decline in the amount of solder used had a somewhat greater impact than the decline in average tin content for both automobiles and light trucks. For the 1950–78 interval, the reverse was true; the decline in the average tin content of solder contributed more to the level of decline in tin usage than the decrease in solder per unit. For all its potential effect, the non-soldered aluminum radiator has had little impact on the level of solder and tin in solder usage.

Several underlying factors are responsible for the changes in those apparent determinants—solder use per radiator and the average tin content of the solder—that have resulted in substantial reductions in tin use for radiator soldering. Prior to World War II, the dip-soldered cellular radiator was commonly used for automobiles and light trucks. During the war, in response to limited tin supplies and government regulations, the tin content of radiator dipping solder was halved. The sweat-soldered radiator, substituted for the dip-soldered radiator in the late 1940s and early 1950s, utilized substantially less solder than a comparable dip-soldered radiator, but required solder with tin contents akin to prewar levels. The net effect was to slightly increase per unit usage of tin in the early 1950s over immediate postwar models, though substantially less tin was used than for prewar radiators.

Concurrently, copper shortages during the Korean War and the normal objective of reducing the cost of materials resulted in the reduction of the size of the average radiator. To maintain heat-dissipating capacity, the operating temperature and pressure of the radiator were raised. Lower tin content solders, which have a higher melting point and are stronger under sustained loads, were consequently adopted in the mid- to late 1950s for the tubes-to-header joint. While the impetus for this change in solder composition was a reduction in copper-brass input per given level of heat rejection, it produced a concomitant reduction in solder material cost. In the 1960s and 1970s the tin content of tube coating, and to a lesser extent tank-to-header solders, also declined, primarily to maintain the proper heat balance during the sequential soldering operations, but also due to material cost savings.

In the latter half of the 1970s the amount of solder used per automobile radiator declined as the output share of smaller cars increased considerably. With light trucks, average per unit solder usage increased over this period as the market share of the larger class of light trucks gained markedly.

All indications point to a further decline in tin input per radiator in the future. Even if the welded or plastic tank is not adopted, the average tin content of radiator solders is likely to drop, perhaps to less than 5 percent compared to the 1978 average of 16 percent, in response to the escalation of tin prices. The continuation of the four-decade-long battle between the copper-brass and aluminum radiator and the downsizing of the automobile ensures that further efforts will be made to reduce the material inputs of copper-brass, solder, and tin in solder.

Automobile Body Solder

The body shell of an automobile is constructed from individual sheet metal pressings which are welded together and to the body frame. Of the many feet of weld seam necessary to join the shell panels and frame, a few are visible on the exterior of the coach. Body solder is used to mask over such traces of the assembly operation, to give the coach a smoothly contoured exterior, and to prevent corrosion at the joint (or weldment). Unlike other end uses for solder, body solder serves as a void filler and is not required for the physical joining of metal components, though its presence does add some strength to welds. Tin-lead solder has a pasty consistency long enough to permit contouring, strong adhesion to the steel panels, neutrality with respect to the body paint, and dimensional stability through repeated changes in temperatures over the vehicle life-span.

There are two distinct stages in body soldering. The sheet metal must be first tinned with a thin film of tin-lead alloy. The film must completely and permanently bond to the sheet metal surface, for it provides the base to which the filler metal adheres. The pasty filler metal is then trowelled into the void and worked into the contour desired. Once solidified, the filled area is brought to a flush, smooth finish with abrasive sanders. The vehicle body is then ready for priming and painting.

While tin-lead alloys adequately serve as body filler, their use does pose several disadvantages. Soldering requires a higher level of labor skills and judgment than most other production line operations. The molten alloy is applied to the body metal by hand; inattention in application may result in metal warpage or dangerous metal spattering. The grinding operation creates lead dust which poses a significant health hazard. For decades the task has been performed in expensive enclosed booths with the workers donning respirators. The firms have also conducted costly biological monitoring of the affected workers. Furthermore, with a specific gravity of 10.6 (.382 pound per cubic inch), body solder does not have a particularly favorable volume-to-weight ratio for a material whose primary function is void filling.

Body Solder for Repairs

This section focuses on the use of body solder for the tinning and filling of new car design welds, which excludes the use of body solder for repair purposes. Since the latter has been an important and interesting application, it is briefly considered here.

Dents, creases, punctures, and poor matchups of body sheet metal are common occurrences on an automobile production line. Fenders, doors, and hoods are especially vulnerable to such incidental damage. Solder is often necessary to cover over these flaws. Unfortunately, data on the incidence of new car repair soldering, and on the average amount of solder used per incident, are unavailable. Executives from different firms reported tin contents of solders currently used for body repair ranging from 1 to 20 percent. A production-line manager relates that "some time ago," 35 percent tin solder was used but its high cost promoted the adoption of lower tin contents. Without more information, no estimate of the materials consumed in production-line repair soldering can be made.

Once an automobile leaves the showroom, its sleekly contoured body is subject to a fair likelihood of injury. In the past, body shop repairmen used tin-lead solders to restore the damaged metal to approximate original contour. Due to the diffusion of plastic fillers, this end use for tin-lead solders has disappeared except for a minor demand supported by restoration purists and the few repairmen who "feel more comfortable" with metal fillers.

According to trade press articles, 20 to 40 percent tin solders were used for body shop repair around World War II. No estimates are available, though, on the amount of solder used. Obviously, usage per restoration would vary greatly. One fact that is certain, however, is the immense population of damaged automobiles and trucks requiring the use of body filler. Total registrations of passenger cars, trucks, and buses increased from 50 million in 1950 to nearly 150 million in 1978. Each year, a phenomenal 26-28 percent of all passenger vehicles and 17-20 percent of all trucks registered are involved in reported accidents. As a rule of thumb, twice as many motor vehicles are involved in accidents than are produced in any given year.

Assuming for the sake of estimation that one-third of the motor vehicles involved in accidents require 1 pound of 25 percent tin solder (both solder in tinning paste and filler), one finds that tin use in repair solder is roughly comparable to estimates of tin used in production-line soldering. For example, as estimated below, the tinning and filling of weld joints in passenger vehicles required 22.1 thousand tons of solder with 690 tons of contained tin in 1955. That same year, 14.5 million passenger vehicles were involved in accidents, which, given the stated assumptions, would require for repair 2.2 thousand tons of solder containing 545 tons of tin. Repair of the 1.95 million trucks involved in accidents in 1955 would require an additional 270 tons of solder containing 73 tons of tin. Thus during the period 1950-60, it is roughly estimated that 2.2-2.7 thousand short tons of solder with 545-680 tons of contained tin were utilized annually for motor vehicle body repair.

By the mid-1960s, the auto-repair market for tin-lead solder had virtually disappeared due to the substitution of plastic fillers. The first "generation" plastic fillers (or cold solders) were developed during World War II specifically as a tin-solder replacement, and by 1947 were commercialized for auto-body repair. These early plastic fillers did not gain general acceptance due to their low adhesion and poor finishing characteristics. Within a few years, however, new epoxies appeared on the market that provided the characteristics the earlier plastic fillers lacked. By 1955, General Motors was offering to repair shops two-part epoxies suitable for metal and fiberglass body-patching (Chemical Week, 1956, p. 42). In 1962, Ford screened commercially available

plastic body fillers and recommended one brand to their service centers.

Based upon prices quoted in a 1956 trade press article, plastic and tin-lead fillers cost the purchaser about the same per unit volume. The use of epoxies, however, offered a significant total cost advantage as less labor skill and time are required to apply cold plastic fillers compared with partially molten tin-lead solder.

New Car Body Solder Usage Before and During World War II

According to various trade journal articles, the solders utilized prior to World War II for new car bodies ranged from 15 to 50 percent tin, with 20-25 percent the most common. Tinning solder was generally of somewhat higher tin content than solder used for filling (Pratt, 1952, p. 53; *Tin and Its Uses*, 1940, p. 13). Other sources report that each soldered, mass-produced automobile body required one-third to three-quarters of a pound of tin, which translates to a solder usage rate of $1\frac{1}{3}$ to $3\frac{3}{4}$ pounds of filler solder, given an average 20-25 percent tin content, and an allowance for tinning solder. In calculating 1941 body solder material usage (table 3-8), each soldered, mass-produced passenger vehicle body is assumed to have used 1/4 pound of 35 percent tinning solder and 3 pounds of 22.5 percent tin filler solder.[13]

Gillett (1940) states that one of the major automakers (accounting for perhaps 30 percent of U.S. output) did not use body solder at all, pounding out dents and grinding down welds rather than solder-filling them. Since one major car manufacturer was already successfully doing without body solder, and since the need to fill body weldment and rough spots was hardly a priority item, the National Academy of Sciences (1941) suggested that the use of body solder could be eliminated altogether with no hardship imposed. Ten days after Pearl Harbor, War Production Board Order M-43 prohibited the use of tin for automobile solder; three weeks later, WPB Order M-38-C prohibited nonessential uses of lead, of which new car body solder headed the list. These restrictions on body solder were in a sense inconsequential, for passenger automobile production was slashed to near nil as the plant facilities, labor, and materials were diverted to the manufacture of defense-related motor vehicles.

Late in the war, with the prospect of commercial auto production resuming under conditions of restricted tin availability, material engineers investigated low-tin and tinless alternatives for body solder. In the spring of 1946, Chrysler engineers, Houssner and Johnson, reported that after testing

---

[13]In contrast to mass-production models, low-volume, speciality, and custom bodies could require up to 75 pounds of solder. Using labor and material-intensive solder for body contouring was deemed more economical than the fabrication of costly, detailed body panel dies.

Table 3-8. Consumption of Solder and Tin in Solder for New Passenger Automobile Body Tinning and Filling, 1941, 1950–78

			Tinning Solder				Filler Solder				
Year	Automobile Production (thousands)	Proportion Body Soldered	Tinning Solder Consumed per Soldered Vehicle[a,b] (pounds)	Tinning Solder Consumed[c] (tons)	Tin Content (pounds per pound solder)	Tin Consumed (tons)	Filler Solder Consumed per Soldered Vehicle[b] (pounds)	Filler Solder Consumed[c] (tons)	Tin Content (pounds per pound solder)	Tin Consumed (tons)	Total Tin Consumed (tons)
1941	3780	.7	.25	300	.35	105	3	3600	.2250	804	909
1950	6666	1.0	.16	492	.26	128	6	18142	.0250	454	582
1951	5338	1.0	.16	394	.26	103	6	14528	.0250	363	466
1952	4321	1.0	.16	318	.26	83	6	11759	.0250	294	377
1953	6117	1.0	.16	451	.26	117	6	16649	.0250	416	533
1954	5559	1.0	.16	410	.26	106	6	15129	.0250	378	484
1955	7920	1.0	.16	584	.26	152	6	21556	.0250	539	691
1956	5816	1.0	.16	429	.26	112	6	15828	.0250	396	508
1957	6113	1.0	.16	451	.26	117	6	16639	.0250	416	533
1958	4258	1.0	.16	314	.26	82	6	11588	.0250	289	371
1959	5591	1.0	.16	412	.26	107	6	15216	.0250	380	487
1960	6675	1.0	.16	492	.26	128	6	18166	.0250	455	583
1961	5543	1.0	.16	408	.26	106	6	15085	.0250	377	483
1962	6933	1.0	.16	511	.26	132	6	18870	.0250	472	604
1963	7638	1.0	.16	563	.26	146	6	20787	.0250	520	666
1964	7752	1.0	.16	572	.26	149	6	21097	.0250	527	676
1965	9306	1.0	.16	686	.26	179	6	25326	.0250	633	812
1966	8598	1.0	.16	634	.26	165	6	23401	.0250	585	750
1967	7437	1.0	.16	548	.26	142	6	20240	.0250	506	648
1968	8822	1.0	.16	650	.26	169	6	24010	.0250	601	770
1969	8224	1.0	.16	606	.26	158	6	22381	.0250	560	718
1970	6547	1.0	.16	483	.26	125	6	17654	.0250	445	570
1971	8584	.98	.08/.16	479	.26	125	3/6	17972	.0250	449	574
1972	8828	.96	.08/.16	474	.26	123	3/6	17760	.0250	444	567
1973	9667	.96	.08/.16	509	.26	132	3/6	19069	.0250	477	609
1974	7324	.96	.08/.16	388	.26	101	3/6	14543	.0245	356	457
1975	6717	.95	.08/.16	354	.26	92	3/5.5	12519	.0245	307	399
1976	8498	.96	.08/.13	396	.26	103	3/5	15098	.0240	362	465
1977	9214	.97	.08/.13	430	.25	107	3/4.5	15324	.0235	360	467
1978	9177	.96	.08/.10	367	.25	92	3/4	14346	.0230	330	422

Sources: Data on U.S. production of passenger vehicles are from the Motor Vehicle Manufacturers Association, 1979, *Facts and Figures*. Other data are based on information available from the trade press and interviews with industry personnel.

[a]Figures shown in this column indicate the weight amount of solder alloy in tinning paste, except for 1941 when molten solder was directly applied.

[b]When two numbers appear in this column (after 1970), the left hand number indicates tinning or filler solder usage for vinyl-roofed vehicles, which use plastisols and body solder, and the right hand figure indicates tinning or filler solder usage for those vehicles that use body solder exclusively.

[c]Figures in this column assumed that all vinyl-roofed cars used plastisols for the roof to rear quarter joint seam and tin-lead solder for other joints after 1970. The vinyl-roofed proportion of the total passenger vehicle output was: 43 percent for 1971, 46 percent for 1972, 49 percent for 1973, 48 percent for 1974, 47 percent for 1975, 46 percent for 1976, 48 percent for 1977, and 41 percent for 1978.

numerous formulations, two suitable alternatives were found: a tinless solder (4.5-5.0 percent antimony, 0.7-1.2 percent copper, 0.8-1.0 percent silver, and the balance lead), and a low-tin solder (3.4-4.0 tin, 3.4-3.8 antimony, 0.3-0.5 silver, trace of copper, and the balance lead). It was claimed that these solders could be directly applied to steel without tinning. However, silver-bearing solders, if they ever made it to the production line, were undoubtedly a short-lived phenomenon. The industry rapidly adopted the tinning pastes and low-tin body solders commercialized by General Motors' Fisher Body Division in 1947.

New Car Body Solder Usage Since 1950

The amount of solder and tin in solder consumed in the tinning and filling of new automobile body seams has been estimated for the years 1941 and 1950–78 on the basis of information from the trade press and interviews with industry personnel. The results, which exclude body solder used in trucks, are shown in table 3-8. The trend in intensity of tin use in body solder per thousand new automobiles is illustrated in figure 3-17 and can be divided into three periods: a transitional phase between 1941 and 1950, a static

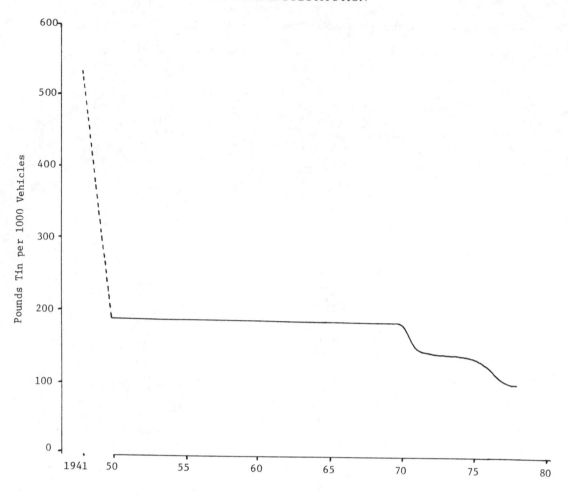

Figure 3–17.   Tin in solder consumption per 1,000 new automobile bodies, 1941, 1950–78 (pounds). [From table 3–9.]

period from 1950 through 1970, and a period of decline since 1970. During the static era, the apparent determinants of tin consumption remained unchanged, except for the number of passenger vehicles produced. This was not the case, however, during the transitional and post-1970 periods, as is described below.

TIN CONTENT OF BODY SOLDER.   This section first examines trends in the tin content of filler solder and then considers the composition of the body solders used for tinning.

1. Filler Solder. As previously mentioned, automotive materials engineers were engaged in seeking low-tin or tin-less body solders late in the war. After some false starts, engineers at General Motors developed alloys containing 2.5 to 4.0 percent tin that would become the commercial standard of filler solder for three decades. Both General Motors and Ford used solders with tin contents at the low end of the range; hence, for the purposes of calculating material usage, it assumed that from 1950 to 1973 the in-

dustry utilized a body solder composed of 2.5 percent tin, 5.0 percent antimony, 0.5 percent arsenic, and 92 percent lead. According to Pratt (1952), a GM materials engineer, the industry found the low-tin solders to yield a satisfactory product; thus there was no impetus to return to the more expensive, higher tin content body solders of the prewar era.

Pratt was of the opinion that 2.0-2.5 percent was the lowest tin content feasible without incurring a reduction in quality. Although this has since been disputed by materials engineers, plant operators and production-line solderers still feel tin is essential to product quality and line productivity, and have resisted all efforts to reduce or eliminate it from body solder. A "compromise" body solder containing 0.8 percent tin exists at General Motors, but is little used.

Ford has been somewhat more successful in adopting the use of minimal tin content solders in response to escalating tin prices. In 1970, the company released a specification for a 1 percent tin content body solder and by 1979 it comprised 50 percent of Ford's total body solder consump-

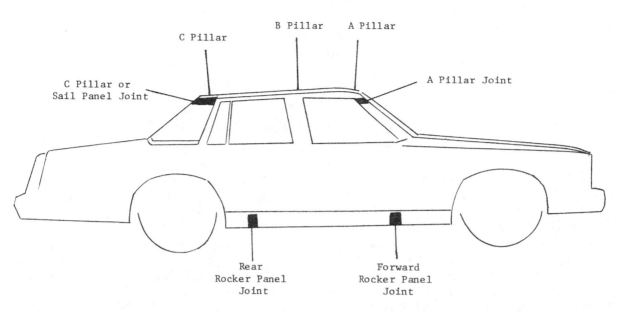

Figure 3–18. Location of body panel joints where body solder is commonly applied.

tion. The remainder was traditional 2.50-2.75 tin solder. Ford production-line operators, like those at General Motors, have expressed dissatisfaction with the 1 percent tin solders. Consequently, the share of reduced tin solders may decline in the future over the short period that body solders are likely to remain in use.

Curiously, the low-tin body solders (2.5 percent tin) so readily adopted by U.S. manufacturers after World War II have made only modest inroads in Britain. In the early postwar period, when tin availability was problematical, an alloy containing 22.5 percent tin plus some antimony was considered by British manufacturers as offering the optimal compromise between cheapness and operational efficiency. Experiments undertaken at that time with lower tin content alloys reportedly proved unsuccessful. With the easing of constraints on tin supplies, the tin content of body solders crept upward. In 1964, the Lead Development Association (LDA) reported that the majority of British automakers utilized solder containing 27-28 percent tin. Perhaps two out of the dozen British auto manufacturers had adopted 15-16 percent tin solders for bulk applications and the standard U.S. body solder with 2.50-2.75 percent tin "where small amounts are to be applied" (Lead Development Association, 1964, p. 6). Low-tin body solders have not won any converts since the LDA investigation, and currently account for less than 7 percent of the filler solder consumed in Britain.

2. Tinning Solder. While solder paste, which is powdered solder in suspension in liquefied flux, has been in limited use at least since the early 1930s, its adoption by the automobile industry after World War II for tinning was the first major mass application. Pratt (1952) reported that the tin in the automotive tinning paste varied from 15 to 30 percent in the early postwar era, with most manufacturers on the high side of this range. General Motors has continued to use 25 percent tin solder in paste since 1947, while Ford has long utilized 28-30 percent tin for this purpose.

Faced with rapidly escalating tin prices, about one-half of the Ford assembly plants switched to the use of 20 percent tin solder pastes in 1977. As with the previously noted change in the tin content of filler solder, some operators reportedly had difficulty in adapting to the use of lower tin content pastes and have argued for a return to 28-30 percent tin in tinning pastes.

BODY SOLDER USE PER SOLDERED VEHICLE. Before examining trends in body solder use per vehicle, it is useful to point out the weld seams where body solder is used on the contemporary passenger vehicle. The junction of the rear quarter and roof panel at the C-post or sail panel, shown in figure 3-18, is the single most important site of body solder use. The sail panel joint may be as long as 20 inches. Other weldments where solder is applied include the junction of the A pillar to the roof panel and the rocker panel to the rear and forward quarters; these joints are typically 3 inches long.

The seams that are solder-filled are resistance-welded lap joints. In cross-section the joint is pan-shaped: The trough of the joint is .20 inch deep and some 1.25 inches wide. The slope of the wall steepens inward, with the overall width of the depression below the contour line being about 2.75 inches wide. By calculating the volume of the weldment trough, assuming a certain combined joint length, and multiplying by the density of body solder, .382

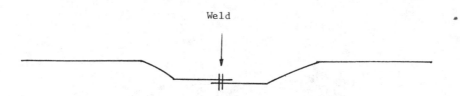

Figure 3–19.   Cross section of a body panel lap joint.

pound per cubic inch (10.6 grams/cm³), the per vehicle use of body solder can be estimated.

During 1950–70 the average passenger vehicle had 4 feet of weld seam solder filled. Given the preceding volume-density relationships and cross-sectional area of a weld seam, the average vehicle over these years required 6 pounds of solder filler and ¼ pound of tinning paste (containing .16 pound of solder), as indicated in table 3-8.[14] Given an estimate for average tin content of filler and tin paste solder, this translates to 190-195 pounds of tin per 1,000 vehicles, as diagrammed in figure 3-17 (of which tinning solder contributes 40-50 pounds).

In the 1970s, redesign of the passenger auto led to a decline in the average per vehicle consumption of solder. First, vinyl plastisol for filling the roof panel to rear quarter joint on vinyl-roofed cars was adopted. Then, to accede to consumer preference and mileage standards, automobiles were downsized and subcompact models introduced, resulting in a shortening of the average length of the roof-to-rear quarter joint. In addition, less expensive models have been introduced, which while still using some solder for design purposes, leave previously solder-filled weld seams unfilled.

1. Plastic Fillers for Vinyl-Roofed Car Bodies. Since the early 1970s, heat-curing plastisols (a dispersion of polyvinyl chloride resin in a plasticizer) have been employed to fill the C-post joint when the roof and rear quarter panel are vinyl covered. Since between 41 and 49 percent of all U.S. passenger cars produced during the 1970s had vinyl roofs, this innovation significantly reduced body solder use. Furthermore, these cars are generally the full-sized models that would be above mean consumers of solder had this change not taken place. A recent full-sized vinyl-roofed car contains some 3 pounds of solder, about a pound less than a compact Ford hardtop.

Vinyl-roofed cars were introduced in 1966 and by 1967 held nearly one-quarter of the passenger car market. For the

first four or five model years, body solder underlay the ''soft-top,'' though undoubtedly production line tests were run on certain models before the swift industry-wide conversion to plastisol fillers in 1970–72.

The plastisol fillers offer a number of advantages over tin-lead solder. No special surface preparation analogous to tinning is required, thus eliminating an operational step. The nontoxic plastisol is puttylike at room temperature and is simply trowelled into the joint and contoured, so there is no metal warpage, spattering of hot metal, or noxious fumes. After heat-curing, the joint is wet-sanded to give the precise contour desired. The plastisol filler provides a good bonding surface for the solvent-type adhesives that are thinly spread over the roof and rear quarter to make the permanent bond with the underside of the vinyl fabric.

Plastisols also offer a cost and weight advantage over tin-lead fillers. Whereas four pounds of body solder may be utilized on a full-sized automobile on roof-to-rear quarter joints, slightly less than one-half pound of plastisol is required. With even crude estimates of material cost, it is obvious that the plastisols offer a significant material cost advantage, as shown in table 3-9. These cost estimates do not include energy, labor, or capital equipment costs. All evidence indicates that if these costs were factored in, the savings of vinyl plastisols over metal body solder would be even greater.

2. Automobile Downsizing and Design Changes. The U.S.-built small car all but disappeared after World War II, with the modest demand for these products met by imports. In the early 1960s, a few of what were then considered ''compact'' models were introduced, such as the Chrysler Valiant, Dodge Dart, and Ford Falcon. The full- and intermediate-sized cars, though, dominated the market.

In the 1970s, particularly the latter half of the decade, the compact and subcompact vehicle class emerged as the

---

[14]These estimates are consistent with published figures. Shea (1950, p. 11), in a survey article on tin, offers a figure of 5 to 10 pounds of solder per vehicle. Pratt (1952, p. 55) of General Motors, in a comprehensive discussion of body solder, presumes an average of 5 pounds per vehicle. A 1970 National Materials Advisory Board report states that design changes have reduced the amount of body solder required for new cars, yet nearly 0.2 pound of tin are required per average car body (National Materials Advisory Board, 1970, p. 29). Given the tin content of tinning and filler solder, this NMAB value corresponds to ¼ pound of tinning paste and 6½ pounds of filler metal.

Table 3-9. Per Vehicle Material Costs of Fill per Roof Panel to Rear Quarter Joint

(cents per vehicle)

	Plastisols	Body Solder		
		Paste	Filler	Total
1970	7	7	79	86
1976	13	14	128	142

Source: Estimated from trade journal articles.

U.S. industry growth sector. Led by General Motors, full- and intermediate-sized models also have been downsized in reluctant response to often fickle consumer demand and strict Federal standards under the Energy Conservation Act of 1975, which required enhanced fleet fuel economy.

The principal impact of this general trend on solder usage is the reduction in the average length of roof panel to rear quarter joint. On full-sized and many intermediate class vehicles, the trapezoidally shaped rear quarter side panel or sail panel has been shortened from its maximum extent of 18-20 inches to 12-14 inches. With the growth in market share of smaller automobiles, the average length of the roof panel to C-post joint declines. Some intermediate class vehicles and the squareback compacts employ a rectangular C-post, which may have a seam of 8 inches. "Fastbacks" and models with wrap-around windows often have a rear roof support rail as narrow as 2 inches.

A compact car with a 6-inch wide C-post and all other design weldments filled as described earlier would require 4 to 5 pounds of solder per vehicle.

3. Elimination of Solder-Filled Joints. A number of General Motors passenger vehicle models have not used solder on the rocker panel to rear and forward panels for a number of years. Solder was replaced with a film of sealant and a metal plate cover.[15] In the mid- to late 1970s, unsoldered C-pillar joints first appeared on the Pontiac Astre, Pontiac Sunbird, Buick Skyhawk, and Dodge Aspen. While these models apparently used solder for other design purposes, the elimination of C-pillar joint solder filling helped to reduce costs on these less profitable lines of automobiles. This increase in unsoldered joints, coupled with the decline in automobile size already examined, produced an estimated reduction in average solder consumption from 6 to 4 pounds per hardtop vehicle.

NONSOLDERED AUTOMOBILES. Pre-World War II consumption of body solder per soldered, mass-produced motor vehicle ranged from 1 to 4 pounds, with a substantial percentage of vehicles, perhaps 30 percent, reportedly not using any body solder at all. In the years immediately after World War II, body solder usage patterns changed markedly. Consumer preferences for comfort and styling were satisfied with polished, amply proportional, more expensive exteriors. Buffed-out weldments were no longer acceptable. Consequently, solder use was extended to virtually all passenger cars and per vehicle consumption of solder increased over prewar levels.

During the period 1950–70, all domestic, mass-produced automobiles utilized body solder.[16] Since then, less expensive subcompact and compact models have been introduced that do not use solder for design purposes.

The first model not to use body solder was General Motors' Vega. This hatchback vehicle had a very narrow rear roof support pillar and at the base of this pillar a 2-inch long open joint. Redesign of the roof has hidden the A-pillar to roof joint below the hood panel. The elimination of design body solder has been passed on to the modified Vega descendant, the Monza Coupe, and to the Chevette, which were both introduced in 1975. Thus, in 1975 over 5 percent of total U.S. automobile output did not use body solder. General Motors' X-car series, Phoenix, Skylark, and Omega, also do not use body solder, but since the series did not go into production until 1979, the X-car is not a component of the numerical analysis in table 3-8, which terminates with 1978.[17]

It has been an economic decision to leave joints open. Elimination of the materials, labor, and expensive hardware associated with solder application and finishing reduces cost on a class of vehicles that have a narrower profit margin than full-sized vehicles. It is considered that customers for subcompacts desire comparatively low-cost, fuel-efficient transportation and are less likely to respond adversely to minor cosmetic deficiencies than the customers for intermediate- and full-sized vehicles where sleek styling and elegance are marketed qualities.

Government Regulation and the Future Use of Body Solder

By 1985–87, it is possible that tin lead alloys will no longer be used on mass-produced automobile bodies as filler. While the decision to do away with body solder will be economic in the sense that automobile manufacturers will find it more profitable to curtail than to continue solder usage, the necessity to make the decision is being imposed by the government. Specifically, the total costs of compliance with federal regulatory standards for occupational exposure to lead may be so high that lead body solder usage by the major auto manufacturers will be terminated.

Basically, the Occupational Safety and Health Administration (OSHA) ordered that the occupational exposure level to airborne lead be lowered through the use of active engineering systems to 50 micrograms of lead per cubic meter of air, and that an extensive program of biological monitoring of exposed workers be instituted. While the affected industries, including automobile manufacturers, challenged the regulations, OSHA's requirements have been upheld by the Supreme Court.

OSHA has eased the compliance schedules by granting variances, but experts inside and outside the auto industry believe that body solder usage will be eliminated by the

[15]The consequent reduction in the extent of solder-filled weld seams has been factored into the previously stated estimate of an average 48 inches of solder-filled seam and 6 pounds of solder per vehicle.

[16]A limited number of specialty sports cars were produced with a molded fiberglass shell for a body. These cars did not use body solder, but comprised substantially less than one-half percent of annual output and are thus not indicated on table 3-8.

[17]Also not included in the numerical analysis is the Ford Fiesta, another subcompact which also does not use body solder for design purposes. The Fiesta, in production since 1976, is Ford's "World Car." It is assembled in southern Europe and exported to the United States and other markets.

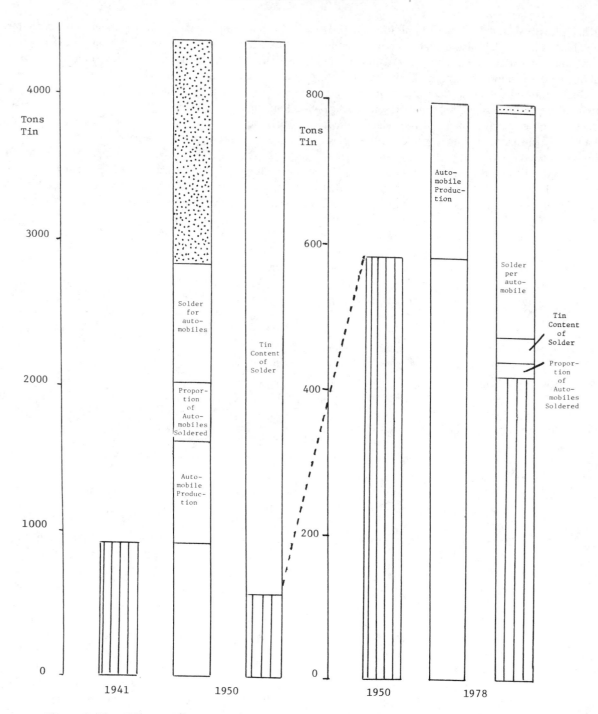

Figure 3–20. Effects of changes in the apparent determinants on tin consumption for automobile tinning and filling.

mid-1980s as a result of these regulations. Alternatives to tin-lead body solders are being developed.

The leading candidate to replace body solders is a silcon bronze alloy applied by welding. General Motors has utilized this method on the J-car model line, introduced in 1981, and Ford has a development program underway. The welding technique requires that the gap between the body panel to be filled by the silcon bronze compound be only a quarter-inch wide, far narrower than with soldering. Higher quality, more precise, sheet metal body panel stampings are required.

It has been suggested that joint seams could be left open, as has been the case with some General Motors compacts. Longer joint troughs would be inexpensively plugged with an ornamental strip. This solution, however, does not appear consistent with the objective of U.S. automakers to upgrade the quality of their products' "fits and finishes" to meet the high standards of the imports.

The automakers would like to use plastic fillers, as is now the case for low-volume production, body shop repair, and vinyl-roofed C-post joints. Indeed, they have spent millions of dollars over the past three decades with precisely this objective, but so far with little success. The principal difficulty is that the lengthy cure time required (20 minutes) is incompatible with a high-speed, mass production-line sequence of operation. Plastic fillers are also not as resistant to the natural elements as desired, nor do they exhibit good paint-over characteristics. Barring a developmental breakthrough, it does not appear likely that they will be adopted for a mass-production body manufacture.

Having lost the judicial challenge to OSHA's lead standard, the Lead Industries Association has launched a development program to fully robotize solder grinding, the auto body soldering operation which generates the highest concentrations of airborne lead dust. Should this program prove successful and the technology cost-effective in comparison with silcon bronze welding, already in the process of commercialization, and should measures be developed to reduce airborne lead concentration below the limit in the tinning and filler application operation, tin lead body solder usage may be retained.

Comparative Effects of Apparent Determinants

In figure 3-20, the comparative effects of changes in the apparent determinants in consumption for body solder are illustrated for 1941, 1950, and 1978.

Three apparent determinants acted to increase the amount of tin used in body solder between 1941 and 1950. First, the number of automobiles produced increased by 76 percent. Second, per vehicle use of filler solder doubled. Third, the use of body solder was extended to virtually all passenger vehicles from an estimated 70 percent in 1941. These last two changes reflected postwar consumer preference for a higher quality of body finish.

Reducing tin consumption was the substitution of low (2.5 percent) tin solders for the 20-25 percent tin solders used in 1941. The low-tin solders were developed in response to concerns for the security of tin supply, or more directly, in anticipation of an extended period of government allocation of available tin. They were adopted readily because they offered significant material cost savings with no reduction of product quality. Of comparatively minor importance, the amount of tin required for body tinning was reduced by the use of smaller amounts of lower tin solder.

Between 1950 and 1978, the only apparent determinant acting to increase the amount of tin use was the 38 percent rise in new car sales. The other determinants all acted to

reduce consumption. As their combined effect outweighed that of automobile production, tin usage in body solder fell over this period.

The adoption of plastisols and auto downsizing have been the primary reasons for the decreased amount of solder used per soldered automobile. Between 1950 and 1978 these have been far more significant in reducing tin consumption than the declining tin content of tinning and filling solder, or the introduction of nonsoldered passenger vehicles.

The adoption of plastisols to fill the roof seam joint on vinyl-roofed cars reduced solder consumption on these vehicles by 3 pounds of filler and 1/8 pound of paste. Most of the properties that tin-lead solder provides for exposed body panel joints are not required for this fabric-covered joint. Since plastisols provide the requisite properties at greatly reduced material cost and with less skilled labor, they were quickly substituted for tin-lead solders soon after the introduction of the vinyl-roofed car. The level of demand for the end product, the vinyl-roofed car, is functionally unrelated to the consequent cost reductions.

Of slightly smaller impact than the adoption of plastisols has been the decrease in body solder usage realized from the declining size of the average fleet vehicle, or more precisely, in the length of the roof joint seam. Obviously, this development is unrelated to soldering costs; rather, it is a response to requirements for increased fuel efficiency and consumer preference. On the other hand, the introduction of less expensive car models with unfilled coach joints is a function of total soldering costs, including labor, capital, material, and energy costs. The cost of tin has been the sole factor prompting Ford's decision to reduce the tin content of tinning and filling solders.

The future of tin in body soldering appears bleak, at best. Compliance with proposed health and safety regulations will significantly increase the capital and perhaps labor costs of soldering, and within five years tin-lead solder may no longer be used as body fill in the production of new automobiles.

Solder Usage in Electronics and Plumbing

The studies of solder usage in previous sections are from the metal can and motor vehicle industries. While these industries are major users of solder, so are the electrical and electronics, coatings, and plumbing industries, as table 3-1 indicates. This section examines in a qualitative manner trends in solder usage in electronic and plumbing applications.

Electronics

The largest consumers of tin in solder are the myriad electrical and electronic applications, currently accounting for

45-50 percent of all tin used in tin-lead solder and slightly more than 20 percent of all solder (table 3-1).

While other solder markets are static or in decline, solder consumption in electronic applications has in recent years increased at 5 percent annually. It is therefore important to analyze the electrical and electronic market, yet it is difficult to do so in other than a general, descriptive manner because thousands of specific products are involved and the technology of soldering is far more complex than for the end uses previously described.

The market may be broken down into the electrical (including solder for cable splicing and light bulb stems) and the electronic sectors. The latter encompasses computer and communications equipment, for both military and commercial systems, as well as consumer products, such as radios, television sets, calculators, digital watches, and automotive electronics. This section focuses on the electronic sector, and in particular on consumer entertainment equipment. Consumer electronics account for over 50 percent of the solder used in the entire electrical-electronic market, and entertainment products are the largest subsector of consumer electronics.

Prior to the adoption of the printed circuit board in the early 1950s, the components of a radio or television were mounted onto the chassis and connected to the other components by insulated copper wires. Each wire end was wrapped around the components terminal and the connection was soldered using wire solder and a hand-held soldering iron. Typically, a 40 percent tin solder was used for this task, but for some products 60-65 percent tin solder was employed. Assuming that each solder connection was made with a simple spherical solder drop from a solder wire $\frac{1}{16}$ inch in diameter, about 2 grams of solder with less than 1 gram contained tin would suffice per 100 joints.

In the late 1940s, consumer demand for radios and the newly introduced television was extremely strong. Producers, though, encountered bottlenecks with the wrapping and soldering task, for it took a skilled operator about 5 seconds per connection. Since a typical radio had perhaps 100 connections and a TV set 500, the wrapping and soldering operation alone took $8\frac{1}{3}$ minutes per radio and $41\frac{2}{3}$ minutes per TV. The adoption of the printed circuit board revolutionized the manufacture of electronic products, reducing the time necessary to make these connections by nearly two orders of magnitude.

The printed circuit board consists of a thin film of copper bonded to an insulating plastic board. The wiring (or conductor) pattern is then screened onto the foil with an etch-resistant ink. When the board is immersed in an etch bath, all the copper not inked with the conductor pattern is etched away. Holes are punched through the board at the conductor terminals and wires from the electronic components, which are mounted on the side of the board opposite the conductor pattern, are inserted through the holes linking the component

with the conductor pattern. The board with the copper conductor pattern and the slightly protruding wires face down is then passed over a molten solder bath. The solder adheres to the conductor pattern and the component terminal wires and infills the eyelets, thus making all the component connections in a single operation. Instead of 5 seconds per hand-soldered connection, a board with a hundred connections could be reliably soldered in 10-15 seconds. By adopting this automated soldering technique, one radio manufacturer reportedly dispensed with 99 percent of their production-line personnel while maintaining the same level of output. With the conductor pattern as well as connections soldered, solder use per electronic product increased five to ten times over hand-soldering usage rates.

While some electronic firms used a 50 percent tin solder bath in the early to mid 1950s, 60-65 percent tin solder was soon recognized as vital to product quality. One factor was the need to conduct the soldering operation at the lowest possible temperature lest the plastic board, the adhesive bonding copper to board, or the electronic components be damaged by the heat. An alloy's melting temperature is lowest at its eutectic composition, which in the case of binary tin-lead solder is 63 percent tin and 37 percent lead. Fluidity and wetability, crucial properties in this automated production process, are also optimal at the eutectic composition.

Automated electronic soldering requires that even minor amounts of impurities, such as antimony, zinc, aluminum, or copper, be avoided. Virtually all solder supplied for automated soldering is thus made of primary tin. As impurities are introduced by the soldering operation, the solder bath is changed frequently. The old bath is returned to solder suppliers who refine the material into lower grades of solder suitable for less exacting work, such as radiator soldering.

In 1975, a leading producer of electronic solders introduced as economical substitutes for 60-63 percent tin solder two new alloys with 52-55 percent tin, 2-3 percent antimony, balance lead. These proprietary substitute solders have generated great controversy within the industry. Competitors charge that the presence of antimony and the reduction of tin content retards the wetting action of the solder. Thus, they contend, board coverage is not as good with 60-63 percent tin solders, yet more solder is used, resulting in a reduction in quality with no significant cost savings.

Under the federal specification code, these substitute electronic solders are not approved for work under government contract. Firms engaged in highly sophisticated electronic technology tend to follow the federal specifications regardless of whether or not they are working under government contracts. Consequently, consumer electronics is the only market for the tin-lead-antimony substitute solders. Within the United States, they have captured only a modest market share, with consumption on the order of 1,000 tons annually. Reportedly, these substitute solders have a considerably larger share of the electronic solder market among the offshore production facilities of American-based multinationals, such as General Electric and Zenith.[18]

A second major technological development affecting solder usage is the miniaturization of sophisticated electronic products over the past thirty years. The first computer, ENIAC, which occupied two floors of a building, and probably required some 30 pounds of solder, can now be outperformed by a programmable hand-held calculator. With consumer electronic products, such as the television and radio, the exterior bulk of the product has not changed as drastically owing to the spatial requirements of the picture tube and speakers. The printed circuit board and the components on it, though, have been reduced in size, with a concurrent reduction in solder input per unit.

The transition from hand to automated soldering in the early 1950s was a reorganization of a production process, but the electronic components remained the same. When transistors came into widespread use in the mid-1950s, replacing most of the functions of vacuum tubes, the size of the printed circuit board, hence solder input, did not change substantially at first. A typical radio printed circuit board in the mid-1950s was about 15 square inches. In the early 1960s, however, when the integrated circuit was adopted and the radio became fully transistorized, the areal extent of the circuit board declined to about 4-5 square inches. During the 1960s and 1970s, the number of transistors within an integrated circuit grew exponentially, yet the size of the printed circuit board declined only about 30 percent as additional components were added to improve reception, tone control, and the like. Very crudely, it is estimated that a contemporary, efficiently produced radio or TV probably utilizes one-fifth as much solder as the first models produced with automated soldering technology, and perhaps one or two times as much solder as a 1950 hand-soldered model.

Some individuals within the electronic solder industry maintain that within the next decade, consumer electronic products will adopt the soldering techniques currently employed in the computer and aerospace industries. These involve extremely precise control of solder use and placement. Instead of passing a printed circuit board over a molten solder bath, minute beads of solder in paste can be screened onto a circuit board to serve as the conductor path, or solder preforms of predesigned size and configuration can be positioned precisely at the connection, thus eliminating solder coverage of the conductor path. While the tonnages of tin-lead solder utilized for electronics will decline significantly if pastes and preforms are widely adopted for mass electronics applications, a greater proportion of

the solder manufacturers' costs will consist of value-added rather than material cost.

The technological development of automated printed circuit board soldering for electronic products has profoundly affected solder and tin in solder use. With the initial adoption of this new production technology, solder consumption increased five- to tenfold over hand-soldering consumption levels. Subsequent miniaturization of electronic components and the printed circuit board of typical consumer electronic products has reduced the intensity of solder use one-fifth from these initial levels. The technical requirements of automated soldering necessitate the use of high tin solder; 60-65 percent tin is by far the most common solder employed, though 52-53 percent tin plus antimony substitutes have enjoyed some acceptance in the consumer electronic market.

In contrast to virtually all other markets, electronics has been a strong growth sector for solder in the postwar period. Potentially, though, another episode of profound technological change is possible in the production process of consumer electronics, eliminating the current, solder-intensive bulk application techniques.

Plumbing

Plumbing pipe joining is the oldest application of tin solders. Soldered copper tube joints have been dated to 1800 B.C., while lead pipe joints have been traced to Roman antiquity.

In modern times, four plumbing solders have been in use. Of primary importance have been the 30-40 percent tin, balance lead solders used for joining lead pipe, and the 50 percent tin, 50 percent lead solders used for copper pipe joints. Of secondary importance, solders containing 95 percent tin and 5 percent antimony are used for high-pressure, high-temperature copper pipe joints. While not yet a significant factor in the U.S. market, 95-96.5 percent tin, balance silver solders are used in northern Europe for potable water pipe joints.

Prior to World War II, lead pipe soldered with 30-40 percent tin was the principal plumbing joint. The making of a good lead pipe joint, though, requires a great deal of craftsmanship. It is also a prodigious consumer of solder; 100 lead pipe joints for pipe with a ¾-inch outer diameter would require an estimated 8.5 pounds of solder with 3 pounds of contained tin.

After World War II, copper pipe began to seriously erode lead pipe's share of the plumbing market, primarily because copper pipe took far less skill, time, and materials to solder. While 50 percent tin solder is used on the common copper pipe joint, solder and tin consumption are substantially less than with a lead-wiped joint, on the order of 1 pound of solder and ½ pound tin per 100 pipe joints, for a pipe with a ¾-inch outer diameter. An additional factor favoring copper pipe is its weight; 20 percent lighter than similarly sized lead pipe, it is easier to maneuver. By the mid-1960s, sol-

---

[18] At least two of the major U.S. electronic solder producers have located production facilities in Singapore and Taiwan in order to supply and provide technical service to electronic firms located in the region.

dered lead pipe had been largely driven from the market by copper pipe and nonsoldered alternatives such as plastic pipe, though lead pipe is still in limited use in a few localities.

In the postwar period, it has become a common practice for building engineers to specify the use of 95 percent tin, 5 percent antimony solders for joining copper pipe in high-temperature or high-pressure applications, such as hot water pipes in multistoried commercial buildings, because a high tin-antimony solder offers better creep strength properties than 50 percent tin, 50 percent lead solder. Tin-antimony solder, though, has virtually no pasty range, which makes it far more difficult to work with. A conscientious, experienced plumber can unknowingly make an incomplete joint, and numerous joint failures resulting in substantial property damage have been ascribed to this problem (Sosnin, 1974). As a result of these failures, many plumbing inspectors will not permit its use within their jurisdiction. Currently an estimated 5 to 40 percent of all copper pipe solder is 95 percent tin, 5 percent antimony solder. Solder use per joint is comparable to that of 50 percent tin, 50 percent lead solder, hence tin in solder usage is nearly double.

Since the mid-1970s, the trade press has reported the advantages claimed for the use of 95-96.5 percent tin, balance silver solders for potable water, copper piping joints. In addition to their superior strength properties, their use precludes the possibility of lead contamination of the water from contact with a lead-bearing solder. Proponents also claim that the tin-silver soldered joints require one-third less time and 40 percent less solder than conventional tin-lead solders. Hence their total soldering cost is not significantly higher than that of 50 percent tin solder.

Due to concerns about lead toxicity, the Federal Republic of Germany and the Inter-Scandinavian Ministries of Housing now require that copper tubing water systems for homes use tin-silver solders; British Columbia also recommends their usage. So far, little interest for their adoption has been expressed in the United States.

No yearly estimates of tin or solder usage for pipe soldering have been made. Bureau of Mines data indicate that 715-3,545 tons of tin are currently used for plumbing and other unspecified general metal joining. Information from the Copper Development Association indicates that, while residential and commercial copper tubing output has been volatile, there has been no secular trend of growth or decline over the past two decades. Consequently, tin and solder used in plumbing has apparently remained static since lead pipe soldering became obsolete in the mid-1960s. With the substitution of copper for lead pipe in the early postwar period, the intensity of tin use per soldered joint declined by one-sixth to one-third. In the future, greater use of very high (95 percent) tin solders may stimulate tin consumption for plumbing solder. However, as the next chapter points out, the growth in market share of nonsoldered, plastic

piping is likely to be substantial, and may more than offset any such tendency.

Tin Consumption in Solder

This chapter has examined a number of end use markets for tin-lead solders. In each case, substantial changes in the pattern of tin consumption have occurred. Some of these changes have been evolutionary and continuous; others revolutionary and discontinuous. Periods of relative stability at times prevail, and marked reversals in the direction of change (toward increasing or decreasing tin in solder consumption) are not uncommon.

These changes can be attributed to fluctuations in four apparent determinants of tin consumption: (1) the production of cans, motor vehicle radiators, automobile bodies, television sets, plumbing pipe joints, or other end products for which solder is used; (2) the proportion of these products that actually use solder; (3) the amount of solder required for each product; and (4) the tin content of the solder used.

The production level of the end product is the sole determinant which varies continuously. As a first approximation, one might expect that output would change in accordance with the growth in U.S. population and per capita income. This, however, has not generally been the case. While the production of evaporated milk cans and lead pipe joints has declined precipitously, fruit and vegetable cans and copper piping have remained relatively static, and electronics, aerosol cans, and beverage cans have grown well in excess of U.S. population and per capita income since 1950.

Changes in the other three apparent determinants, which define the intensity of tin in solder use per unit output, are superimposed upon and may act to accentuate or offset trends in end product output. Often, one of these three determinants acts alone to alter the intensity of tin use, such as the decline in the tin content of evaporated milk can solder, the shifting proportion of soldered versus nonsoldered beverage cans, or the miniaturization of printed circuit boards. When two or more of these apparent determinants change concurrently, they may accent each other, as occurred when the adoption of smaller radiators required both less and lower tin content solders, or to offset each other, as occurred when the switch from lead to copper plumbing pipe resulted in the use of less but higher tin content solder.

The underlying factors responsible for such changes in the apparent determinants are numerous. Their influence is specific to particular applications and time intervals.

The physical shortages and the threat of long-term strategic vulnerability of tin during the World War II and the Korean War era induced the adoption of low-tin alloys in the case of canmakers and auto body solder. Since the substitution of low-tin solders for traditional high-tin solders was cost-efficient, posed no significant technological barriers, and required no innovative design breakthrough, it is not clear why this switch did not occur earlier. Apparently, significant inertia in traditional material usage favored the status quo.

Since the Korean War, even though concerns for the strategic vulnerability of tin supply have subsided, one finds little evidence of "inertia" retarding substitution. Material and total soldering costs, technological change, and more recently, public policy, have induced a net decline in tin input for many products, and furthermore, bode a continuation of this decline.

The escalation in the price of tin has stimulated the evolutionary reduction in the tin content of solders used for evaporated milk cans, and for radiator tube coating and tank-to-header joints. These are automated soldering operations where labor input is minimal. Total soldering costs have been more significant in inducing change in the skill-intensive hand-soldering operations. The high costs of hand soldering electronic components resulted in the adoption of a new production technology, which increased output per man-hour tremendously, but also increased tin and solder use per unit of output. In construction, labor-skill and solder-intensive lead pipe was replaced by less labor and material-intensive copper piping.

Technological changes which have markedly affected solder usage may or may not have been prompted by the desire to do so. The miniaturization of electronic components, particularly in the aerospace, defense, and computer markets, which has greatly reduced solder input per unit output, certainly has not been induced by considerations of solder cost. In contrast, the years spent in the development of thin metal sheet-welding technology for can and auto body seams have at least in part been motivated by the desire to produce a stronger, less material-intensive solderless joint.

Public policy has also had a substantial impact on tin-lead solder usage. Concurrent with the rise of Japanese adventurism in Asia in the mid-1930s, the government launched an evaluation of the strategic vulnerability of tin supplies. A government-industry dialogue regarding substitution potential for tin, including tin solders, prompted industry research and development. Indeed, some government funds were directed toward the development of substitutes. During World War II and the Korean War, the government allocated the supplies of strategic materials and placed a ceiling on the tin content of solders, forcing industry to limit its use.

Government regulation has been a factor in solder usage in recent years as well. Most obvious are FDA and OSHA regulations regarding consumer and occupational exposure to lead. Indeed, concern for lead toxicity may completely eliminate the use of solders in three-piece food cans and auto body filling. Regulation may also affect solder usage very indirectly, but just as markedly. The downsizing of motor vehicles to meet government fuel efficiency standards has reduced per vehicle body and radiator solder use, and may eventually induce automakers to adopt the nonsoldered aluminum radiator. Federal regulation affecting aerosol propellant composition has spurred the adoption of nonsoldered aerosol containers.

4

Tin Chemical Stabilizers and the Pipe Industry

Derek G. Gill

Since 1955 the consumption of tin in chemicals has grown at slightly more than 6 percent a year in the United States. The share of total tin consumption going into chemicals has increased sixfold, from 2 to 12 percent (table 1-1).

With the price of tin in recent years rising rapidly, one might expect the demand for tin in chemicals to fall. The upward trend is all the more curious when compared with consumption in other tin-using sectors. Tin usage in tinplate, solder, and bronze and brass, for example, has declined on average by 1 to 3 percent a year since 1955.

This chapter attempts to determine why tin consumption in chemicals has continued to increase despite the sharp rise in the price of tin, and in contrast to other important tin-using sectors. Since the number of tin chemicals runs into the hundreds and the number of their end uses is even larger (an incomplete list is shown in table 4-1), it is necessary to limit the inquiry. This chapter focuses on tin chemicals used as stabilizers in the production of polyvinyl chloride (PVC) plastic pipe. Stabilizers were chosen because they are an important use of tin that in recent years has grown more rapidly than even tin chemicals as a whole.

In addition, PVC plastics have a wide application, including building and construction, packaging, consumer goods, transportation, and electrical uses. However, the rigid applications have the greatest need for stabilizers. Pipe was chosen because it is the single largest end use of rigid PVC plastic. PVC plastic is used to make water pipe, gas pipe, irrigation pipe, processing pipe, sewer pipe, drain-waste-vent (DWV) pipe, and electrical conduit. This chapter concentrates on water, DWV, and sewer pipe, the three largest applications. Together they consumed over 75 percent of all PVC pipe in 1978.

The scope of the analysis is limited in one more respect. The time period covered includes only the years 1964 through 1978. Before 1964, tin stabilizers were not used in the production of PVC plastic pipe.

The methodology for analyzing tin consumption in each of the three pipe markets examined is similar to that followed in earlier chapters. The amount of tin used in the production of pipe is examined by year over the 1964–78 period. Then the apparent determinants of tin usage are considered, along with the underlying factors causing these apparent determinants to change over time.

The amount of tin used in any year t in the production of pipe depends on five apparent determinants according to the following identity:

$$T_t = a_t\, b_t\, c_t\, d_t\, P_t$$

where: T_t = the amount of tin consumed in year t in the production of water, DWV, or sewer pipe in thousands of pounds.

a_t = the tin content of organotin stabilizers in year t.

b_t = the organotin stabilizer content of PVC pipe compound in year t.

c_t = the share of the plastic pipe market by weight held by PVC plastic in year t.

d_t = the share of the pipe market by weight held by all plastics in year t.

P_t = the amount of water, DWV, or sewer pipe produced in year t, measured in thousands of pounds.

Table 4-1. Major Tin Chemicals and Selected Uses

Tin Chemical		Chemical Formula	Use
Organic[a]	Inorganic		
	Tin Fluoride	$Sn\,F_2$	Toothpaste
	Tin Oxide (Cassiterite)	$Sn\,O_2$	Glass opacifier, ceramic pigment
	Tin Sulfate	$Sn\,So_4$	Tin plating processes
	Tin Sulfide	$Sn\,S_2$	Bronzing agent for treating wood
	Tin Chloride (tin tetra-chloride)	$Sn\,Cl_4$	Weighting silk, ceramics, surface treatment of glass, stabilizing perfumes in toilet soap
Triorganotin compounds		$R_3\,Sn\,X$	Wood preservation, pesticides, insecticides, bactericides, antifouling paints
Diorganotin compounds		$R_2\,Sn\,X_2$	Stabilizers in PVC plastics

Source: U.S. Department of the Interior, Dictionary of Mining, Mineral and Related Terms (1968).

[a]Organic tin chemicals are called organotins. Their organic nature stems from the presence of tin-carbon bonds.

Unfortunately, data on the last two apparent determinants are not available. There are, however, reasons to believe that the production of water, DWV, and sewer pipe closely parallels construction expenditures in the United States. For this reason, it is assumed that:

$$P_t = k\,Q_t$$

where: Q_t = construction expenditures in billions of 1978 dollars in year t.
  k — an unknown constant.

Now, the share of the water, DWV, or sewer pipe markets held by all plastics is given by,

$$d_t = Y_t\,/\,(k\,Q_t)$$

where: Y_t = the amount of plastic water, DWV, or sewer pipe produced in year t, measured in thousands of pounds.

Since information on construction expenditures (Q_t) and plastic pipe production (Y_t) is available, specific values can be ascribed to the variables determining tin consumption in the production of PVC plastic pipe if the basic identity shown above is modified in the following way:

$$T_t = a_t\,b_t\,c_t\,(kd_t)\,(P_t\,/k)$$
$$= a_t\,b_t\,c_t\,(Y_t\,/Q_t)\,Q_t$$

Moreover, although it is not possible to determine the values of d_t and P_t as long as the constant (k) remains unknown, the contribution of these two apparent determinants over

time to changes in the use of tin in the production of PVC pipe can be appraised since they vary directly with changes in $Y_t\,/Q_t$ and Q_t.

Before tin consumption patterns and the contributions of the various apparent determinants are analyzed, however, the next two sections examine the role of tin in PVC plastic pipe production as well as certain relevant aspects of the pipe and tin stabilizer industries. The following three sections then investigate tin usage in the water, DWV, and sewer pipe markets respectively.

The Pipe Industry

This section first examines the products the pipe industry manufactures. It then identifies the various materials used to make pipes; considers the technologies involved, particularly in producing plastic pipe; and finally examines the market structure and competitive nature of the industry.

Products and Sectors of the Industry

Pipes, tubes, and culverts are produced today for a variety of applications.[1] The major ones are potable water, drain-waste-vent, sewer and drainage, conduit, irrigation, oil and gas, process, and structural and mechanical. In addition, there are a number of miscellaneous applications.

The functions of these different products are almost self-explanatory. Potable water pipe conveys drinking water, while drain-waste-vent pipe takes care of liquid wastes and ventilation in buildings. Sewer and drainage pipes convey liquid wastes that may include fecal matter. Electrical conduit protects electrical wires, and irrigation pipe is used to convey water from lakes and wells to farm lands. Oil and gas pipes are used to transport oil and gas, while processing pipe transports industrial fluids and effluent. Finally, structural and mechanical pipes are used for high-stress applications in the construction industry.

These diverse applications require different characteristics for satisfactory performance. Before highlighting the requirements for special applications, it is useful to note certain features desirable for all pipe materials. These include resistance to corrosion from rust or chemicals, internal smoothness to reduce flow friction, low coefficient of expansion to minimize expansion and contraction from temperature changes, and easy joining techniques to facilitate installation.

Certain applications also require special features. For potable water transport, it is essential that the pipe material be resistant to stains. Light weight facilitates the installation of drain-waste-vent pipe but contributes to plumbing noise.

[1]The distinction between pipe and tube is not always made clear. In this study, a tube is a flexible thin-walled pipe that can be coiled. A culvert is an extremely large diameter pipe, often used for highway drainage.

Sewer applications require resistance to sulfuric acid attacks and joint integrity to prevent leaks. Electrical conduit requires that the material have a high dielectric constant (that is, be a good insulator), but this has a disadvantage if the pipe is buried underground since electromagnetic locating techniques cannot be used. Oil and gas pipe must be able to withstand extremely high pressures, while irrigation pipe should be flexible, light weight, and hence easy to move about. Finally, for structural and mechanical applications, the material used should possess great strength, high impact resistance, and resistance to deformation.

Due to the many products or sectors of the pipe industry, the numerous materials competing in each sector, the different categorizations used by trade associations publishing pipe statistics, and the absence of reliable data on some materials consumed in certain sectors, a comparison of growth trends along sectoral lines is not possible. However, the next section does show aggregate growth trends for those materials for which annual data are available. It also provides a partial sectoral analysis of the plastics pipe industry with which this study is primarily concerned.

Materials

While many materials are used to make pipe, the prominent ones are ductile iron, cast iron soil pipe, asbestos cement, steel, reinforced and prestressed concrete, vitrified clay, copper, aluminum, wood, and plastics. These materials compete with each other in many applications.

Figure 4-1 shows the aggregate consumption trends of three materials—plastics, cast iron soil pipe (CISP), and copper. Other materials are not included either because data are lacking or because they are not used in significant quantities in the three pipe markets examined.

Two conclusions are immediately suggested by this figure. First, plastics have been very successful in penetrating the pipe industry, growing at an average 20 percent annually over the 1965–78 period. This compares with a 2 percent a year decline in the aggregate use of copper and cast iron soil pipe. Second, since real construction expenditures have increased by only 4 percent a year, the growing use of plastics has presumably been stimulated by the substitution of plastic for traditional materials.

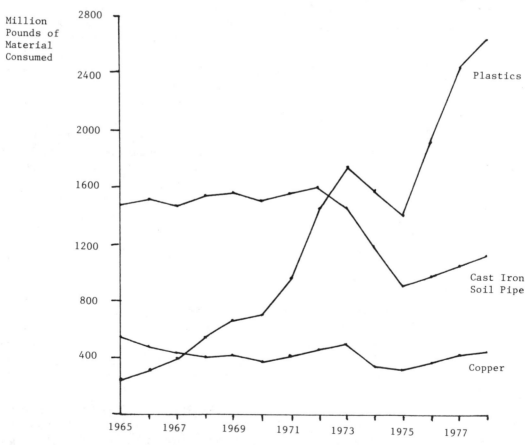

Figure 4–1.   Consumption of plastic cast iron soil pipe, and copper in pipe, 1965–78. [From Plastics Pipe Institute (various years); *Modern Plastics* (various issues); Cast Iron Soil Pipe Institute, *Annual Sales Statistics* (various years); Copper Development Association (1979).]

The consumption of plastic in the manufacture of water pipe, DWV pipe, electrical conduit, sewer pipe, and oil and gas pipe is illustrated in figure 4-2, which shows that plastics find their greatest use in potable water pipe applications. It also indicates that plastic usage has grown at different rates in different sectors since 1971. The fastest growth is recorded for "other" pipe, while the slowest growth is found for water pipe, the largest user of plastics.

Production and Technology

Methods of producing pipe have changed greatly over the centuries. Among the more important factors contributing to this change are (a) the creation of new materials, for example plastics, and (b) input substitution. Naturally, the great physical and chemical differences in the properties of pipe materials imply widely varying production technologies. Some technologies are relatively labor intensive while others, such as cast iron pipe production, are capital intensive. Generally, metal pipe production is more capital intensive than nonmetal pipe production. One exception, however, is clay. Its kiln-drying operation requires a large capital outlay. Because technology has important implications for the substitution process, the subsequent discussion examines the manufacturing technologies of pipe made from different materials, and notes important changes that have taken place over the years.

Plastic pipe is made from many different resins. The most common are polyvinyl chloride (PVC), acrylonitrile-buta-

diene-styrene (ABS), polyethylene (PE), polybutylene (PB), and styrene reinforced plastic or styrene rubber (SRP).

Naturally, the quality of pipe produced depends on the properties of the resin. The mechanical and thermal properties of the more common plastics used to make pipe are identified in appendix 4-1. Though plastics do not possess the great strength of metals, their chemical resistance, flexibility, and light weight give them a significant advantage in many pipe applications.

The manufacturing of PVC plastic pipe, as figure 4-3 illustrates, involves five principal steps. The first step depends on petroleum products from which ethylene or acetylene is derived. The second step is known as the "oxychlorination process," which is simply the combining of ethylene with oxygen and chlorine to form vinyl chloride monomer (VCM), the basic raw material from which PVC is made. The last three steps produce resins, pipe compound, and the final pipe product, and these steps are now discussed in greater detail (Sarvetnick, 1969).

Resin production entails joining monomers[2] of the same type to form homopolymers (or simply polymers) or joining monomers of various types to form copolymers. For instance, vinyl chloride monomer (VCM) can be polymerized[3] to form polyvinyl chloride, or it can be polymerized with a different monomer such as vinyl acetate to form a vinyl

---

[2]A monomer is the simplest repeating structural unit of a polymer.

[3]Mason and Manning (1945) define polymerization as a type of organic reaction in which a complex molecule of high molecular weight is produced from many simple molecules.

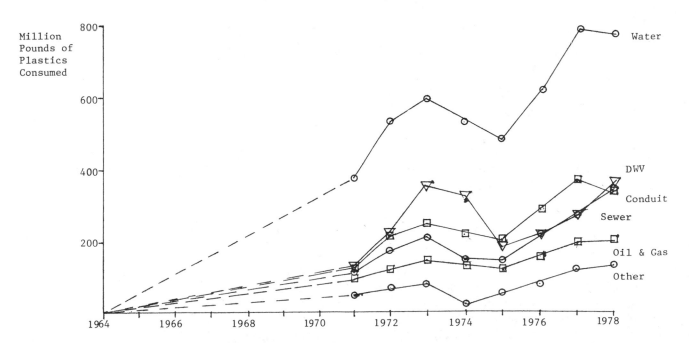

Figure 4–2. The consumption of plastics in various sectors of the pipe industry, 1964, 1971–78. [From Plastics Pipe Institute (various years).]

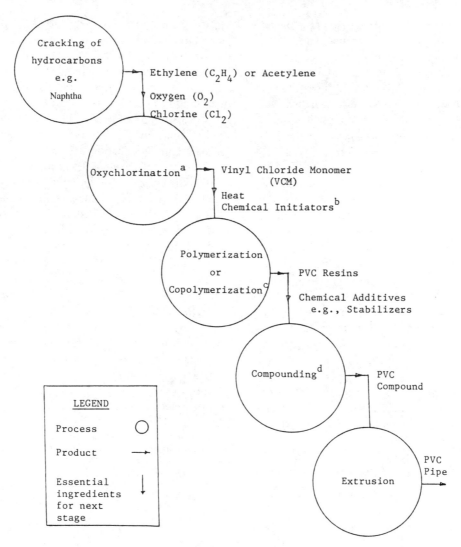

Figure 4–3. Steps in the manufacture of PVC pipe. [a]The oxychlorination of ethylene is represented by the following chemical reaction:

$$\tfrac{1}{2}\,O_2\ +\ Cl_2\qquad +\ 2\,C_2H_4\quad \rightarrow\qquad 2\,C_2H_3Cl\ +\ H_2O$$
$$\text{(oxygen) (chlorine)}\qquad \text{(ethylene)}\qquad \text{(VCM)}\quad \text{(water)}$$

[b]Chemical initiators are just like catalysts—they initiate and control the speed of the reaction and enable it to go to completion. [c]Polymerization or copolymerization is conducted in a watery medium to which suspending agents are added. Emulsifying agents which affect the surface characteristics of the resins are also added. [d]Compounding combines the resins and additives into a uniform or homogeneous state suitable for processing. A comprehensive list of additives used in plastic pipe production is provided in appendix 2.

copolymer. This combining process, which leads to polymers or copolymers, is done in a reactor under stringent temperature and pressure conditions, and requires the use of various catalysts. Different resins, such as PVC and ABS, are produced by combining various monomers, employing different combining processes (Stanford Research Institute, 1975).

Compounding involves mixing the resin with various chemicals to form a compound that is suitable for extrusion. Appendix 4-2 provides a complete list of chemical additives. Among the more common ingredients are lubricants which prevent the mix from sticking to the sides of the extruder, flame retardants which reduce flammability, and impact modifiers or fillers which increase resilience. The mixing of these processing aids with the resin is done in a high-speed mixer to ensure that the resin absorbs the additives.

An additional ingredient that must be added to PVC resins is a stabilizer since these resins are vulnerable to heat and ultraviolet (uv) light. As Sarvetnick (1969) notes, "PVC is unprocessable without the addition of certain compound ingredients such as heat stabilizers. . . ."

Two types of degradation are recognized in the literature: (a) photodegradation,[4] and (b) thermal degradation.[5] As a result, both light and heat stabilization are desirable. The two are quite different; the former guards against oxidation, while the latter provides heat protection.

Throughout this chapter, the emphasis is on heat stabilizers. There are several reasons for this: first, heat stabilizers, depending on composition, can also function as light stabilizers, and their actual classification depends on which function is performed better; second, light stabilization is less crucial since most installed pipe is not exposed; third, and most important, light stabilizers do not contain tin.

Because of poor heat stability, early attempts to process PVC in extruders were not very satisfactory. The high viscosity of the PVC (pipe) compound causes it to flow slowly through the extruder, thus reducing production rates and raising processing costs. Higher extruder temperatures reduce viscosity and increase flowability and output, but they cannot be arbitrarily imposed because of PVC's susceptability to heat. In fact, at 200°F, PVC's thermal degradation begins and its color changes from white to brown, and eventually black (Sarvetnick, 1969). Excess heat causes decomposition as noted before, and loss of desired physical properties, such as flexibility. The stabilizers used most commonly in PVC to inhibit decomposition are organotins or organic tin chemicals, though lead stabilizers and antimony stabilizers are sometimes employed (personal interview).

Finally, extrusion involves melting the free-flowing powdery compound (resin and additives) and forcing it through an orifice. The dimension and shape of the orifice determine pipe diameter and thickness. The main components of an extruder are (a) the hopper, (b) the barrel and the screw, and (c) the die. The pipe compound is poured into the hopper, drawn between the barrel and the screw, and then fed to the die. Pipe is produced by drawing the material through the die.

Because of the susceptibility of PVC to heat, twin and multiple screw extruders were developed. Among their advantages are better temperature control, shorter residence time,[6] and the use of larger cross-sectional dies which allow the manufacture of larger diameter pipe. Single screw machines subject the PVC compound to much higher frictional heat and longer residence time (personal interview). This advancement in extruder technology opened the way for the rapid growth of PVC in the pipe market.

Competing with plastics for the pipe market are clay, concrete, cast iron, and other materials. The manufacturing of clay pipe involves the mixing of different clays to obtain the desired combination, shaping under a minimum pressure of 125 pounds per square inch, then drying, and finally kilning at temperatures of 2,000°F to fuse the clay particles. Kiln-drying is the most costly part of this operation since it is highly energy intensive.

Clay pipe is used extensively for sewer and drainage applications because it is the most chemical-proof pipe. It does not disintegrate and is, "Highly resistant to chemicals or moisture in the soil, chemically active sewage, industrial wastes, abrasive materials in the flow and acids formed by the oxidation of the hydrogen sulphide (H_2S) in the sewer" (Clay Sewer Pipe Association, 1946, p. 10). Because of these properties, clay pipe has a useful life of at least 50 years.

The basic ingredients of concrete pipe are cement, sand, gravel, and water. There are four different ways to produce this pipe. (1) The mix of materials can be placed into different-sized forms where compaction and shaping are accomplished by vibration. (2) The "packerhead" method employs a high-speed rotating ram that forces the mix into a form. (3) The "machine-made tamped pipe" uses a vertically striking tamper, a revolving outer form and fixed inner core. The tamper load can be adjusted to achieve the desired density. (4) Quickly rotating forms achieve compaction by centrifugal force (American Concrete Pipe Association, 1951). Regardless of the manufacturing method, the pipe is then cured by steam spraying or some form of water curing that involves a period of inundation. Concrete's two admirable properties are its high compressive strength and low coefficient of thermal expansion. However, concrete pipe is very heavy and weak in tension. It has a life expectancy of 100 years and is used mainly for sewer and drainage pipe and water mains.

The manufacture of cast iron soil pipe (CISP) first involves the production of pig iron. The molten iron is used to centrifugally cast the pipe in horizontal sand-lined or water-cooled metal molds. This requires that a predetermined amount of molten metal just adequate to make a pipe of specified dimensions be poured into a rotating mold. The remaining steps involve cleaning by sandblasting or tum-

---

[4]Photodegradation means that the polymer begins to degrade (via a process called oxidation) when exposed to light. Light defects include spotting (stains), discoloration (formation of color), hardening (loss of flexibility) and tackiness (development of stickiness). Photodegradation is governed by Grothus Draper's law, which suggests that this action varies directly with the light absorbed or the quantity of luminous energy retained (Chevassus and DeBroutelles, 1963).

[5]Thermal degradation, which results in the breaking up of the long molecular chains of the polymer, is caused by heat. It results in the liberation of hydrogen chloride (HCl), a noxious and corrosive gas and, like photodegradation, causes discoloration.

[6]Residence time refers to the length of time the pipe compound is in the extruder, i.e., from the bottom of the hopper to the tip of the die.

bling machines, inspection and testing, and coating by dipping to give a smooth finish. Cast iron soil pipe finds extensive use in DWV and sewer applications because it makes for quiet plumbing systems and possesses high strength and temperature resistance. Its graphite content also contributes to corrosion resistance (Cast Iron Soil Pipe Institute Annual, 1976).

The manufacture of asbestos-cement pipe first involves the production of asbestos cement by reinforcing a mixture of portland cement with asbestos fibers. Pipe is made in a Magnani machine by depositing slurried asbestos cement by suction on a hollow steel mandrel lined with canvas. Rotating rollers aid in forming the outer profile while compacting the mixture into a dense homogeneous structure. This is followed by two cooling periods lasting a total of 3 to 7 days (United Nations, 1969).

The manufacture of copper pipe starts with a solid-cast cylindrical billet and involves the following five steps: (1) preliminary forming by piercing or extrusion, (2) pointing, (3) cold drawing, annealing and pickling in steps, (4) straightening by a roll, medart, block or hand, and (5) finishing.

Briefly, piercing is done in a Mannesmann machine. The preheated (1,100-1,600°F) billet is forced over a mandrel to form a shell and an arbor—a long steel bar with a high-speed rotating point—does the piercing. Extrusion, similarly, produces a crude-looking tube. After either operation, "pointing" ensues. It is a cold-working operation with a conical point 8-10 inches long, and prepares the tube for drawing. Drawing takes place on drawbenches and reduces outside and inside diameters and wall thickness to specified dimensions. Tube straightening aims to eliminate local or general curvatures and the method chosen depends on size and temper of the finished tube. Finishing involves cutting to appropriate lengths (McMahon, 1965).

Market Structure and Competitiveness

The pipe industry in the United States is highly competitive. Concrete is probably an extreme case. Over 350 small plants make up the concrete pipe industry, each locally owned and managed. Any one firm has only a small share of the regional market, and faces competition from nearby producers.

Strong competition also prevails among the producers of clay, asbestos cement, cast iron, and copper pipes. The plastic pipe segment of the pipe industry is also highly competitive. It comprises some 120 producers, many of whom operate small plants for local distribution.

Most metal pipe producers, prior to the 1960s, produced pipe of only one type of material. The advent of plastics and their rapid growth in the pipe industry, however, led many metal pipe producers to include plastic pipe in their product range. Some copper and cast iron producers merged with or bought out small plastic pipe producers; others constructed their own facilities.

The converse is not true. Plastic pipe producers did not diversify into metal pipe production. Plastic pipe production is not capital intensive. Plastic pipe extruders can be quickly ordered; and since there are many resin suppliers, pipe production can be started in a few months. In contrast, metal pipe production is much more capital intensive, involving the construction of blast furnace facilities in some instances.

Two other aspects of the market structure and behavior of the plastic pipe industry should be noted. First, only a few plastic pipe producers are vertically integrated. Ten plastic pipe producers engaged in PVC resin production and eight produced VCM and chlorine in 1978. Only five were manufacturing their own petrochemical feedstock. So the vast majority of firms just produce pipe. Second, plastic pipe firms have the capacity to use any type of resin, such as PVC, ABS, and polyethylene (PE), and often do. This enables them to compete in many markets since these materials are used with different effectiveness in the various applications. Also, in some applications one material may not be used at all.

Finally, in all sectors of the pipe industry, list prices are seldom published or followed. This is particularly true for the plastic pipe sector.

The Tin Stabilizer Industry

Organotin production officially started in the United States in 1936 when V. Yugve of Union Carbide Corporation was issued a patent for the use of organotin compounds as stabilizers in rigid PVC. Their greater efficiency was immediately recognized. Yet their use in PVC was abandoned, though continued in other applications (Chevassus and DeBroutelles, 1963).

As late as 1957, organotin research was still proceeding at a snail's pace. Despite ample tin supplies caused by reduced U.S. government stockpiling and increased production by tin producers, organotin research was deterred by spiralling tin prices, which came about because the crisis in the Suez area led to a closing of the Suez Canal and increased transportation costs. Further inhibiting their development were toxicity concerns, and the availability of a host of cheap non-tin stabilizers. Among the latter were compounds of barium, cadmium, lead, calcium, zinc, magnesium and nitrogen, as well as some catalysts and other proprietary compounds. However, Metal and Thermit Corporation of New Jersey owned a detinning operation that yielded low-cost tin (*Chemical Week*, July 6, 1957, p. 51). It became the leading researcher and producer of organotin chemicals, and still is today.

Over the years, expiration of patents, high demand for rigid PVC and stabilizers, and the less esoteric nature of organotin chemistry encouraged entry in stabilizer production. Today, there are about forty producers of heat stabilizers, though only eight of them produce *tin* stabilizers for rigid PVC pipe. Of these eight, as table 4-2 shows, the top two control about 70 percent of the market. They do, however, face competition from producers of such substitutes as antimony stabilizers.

Water Pipe

Three types of plastic water pipe are typically distinguished. Main pipe refers to the municipal or rural water main or public works line. It is common in 6-12 inch diameters. Service pipe is the pipe extending from the main line to the first stopcock in the building. This pipe is usually 4-10 inches in diameter. Distribution pipe covers all internal plumbing that distributes water to various usage points. Normally, it is 6 inches or less in diameter. Figure 4-4 illustrates the relationship among these three kinds of pipe. While pipe used for irrigation is not considered, pipe for industrial and commercial uses is.

Trends in Tin Consumption

Annual tin consumption in water pipe over the 1964–78 period is illustrated in figure 4-5. The dashed line represents a period for which reliable data are unavailable. It is known that 1964 was the first year that tin was used in PVC water pipe, and that only minute amounts were consumed. So for all practical purposes, tin consumption for 1964 can be taken as zero. The amount of tin used in water pipe reached a maximum in 1972 when consumption peaked at 950,000 pounds.

Figure 4-6 shows the intensity of tin use in water pipe over the same period. Intensity is measured in pounds of tin per billion dollars of construction expenditures. The latter, in constant 1978 dollars, serve as a proxy for total water pipe consumption. Maximum intensity of use again occurred in 1972 at 4,400 pounds of tin per billion dollars of construction.

Table 4-2. U.S. Organotin Stabilizer Producers

Company	Market Share, percent[a]
M & T Chemicals Inc.	35
Carstab Chemicals Inc.[b]	35
Argus Chemical Corp.	10
Cardinal Chemical Co.	10
Tenneco Chemicals Inc.	
Synthetic Products Co.	
Interstab Chemicals Inc.	10
Ferro Corp.	

Source: Communication with industry.
[a]Estimates based on information from industry sources.
[b]Formerly Cincinnati Milacron Chemicals Inc.

Two distinct trends are apparent from figures 4-5 and 4-6. From 1964 to 1972, tin use grew considerably, climaxing with a 118 percent jump in 1972 alone. Intensity of use also rose rapidly during this period, increasing 105 percent in 1972. In 1973, both tin use and intensity of use exhibited sharp declines, falling 36 percent, while construction expenditures essentially remained unchanged. It is clear that the decline in intensity of use was caused by the decline in tin consumption, but the reasons for the latter are less obvious and will be examined later. Since 1973, more modest declines in both tin use and intensity of use have continued, a trend reversed in 1976 and 1977.

The apparent determinants of these trends in tin usage were identified earlier, and include changes in the tin content of the organotin stabilizer (a_t), the organotin stabilizer content of PVC pipe compound (b_t), PVC's share by weight of the plastic water pipe market (c_t), plastic pipe's share by weight of the total water pipe market (d_t), and the total production of water pipe (P_t). Table 4-3 indicates how these apparent determinants have changed in recent years.

Since the actual value of plastics' share of the water pipe market is not available, changes in this apparent determinant are calculated on the basis of changes in the ratio of plastic water pipe production to construction expenditures (Y_t/Q_t). This ratio, which is shown in table 4-3, closely approximates plastics' share of the total water pipe market multiplied by some unknown constant (k) for the reasons discussed earlier. The actual value of total water pipe production is similarly unavailable, and changes in construction expenditures (Q_t), also given in table 4-3, are used to provide estimates of total water pipe production divided by the unknown constant (k).

The next section examines changes over time in the tin content of organotin stabilizers. Subsequent sections then consider each of the other apparent determinants in turn, analyzing how they have changed and the major underlying factors responsible.

Tin Content of Organotin Stabilizers

The percentage of tin used in stabilizers has exhibited a stepwise decline from 1965 to 1978, as illustrated in figure 4-7. Until 1969, dibutyltin stabilizers averaging 19 percent tin by weight were used extensively. Octyltin stabilizers were introduced in 1965 but their use did not affect tin content, which remained at 19 percent. Stabilizers known as methyltins with a tin content of 15 percent appeared in 1969. Tin usage in stabilizers fell further in 1975 when improved dibutyltin and dimethyltin stabilizers and "Estertins" invaded the market. Their tin content was only 10 percent, roughly half the tin content of stabilizers used ten years earlier.

Two final points should be noted about determinant a_t, the tin content of organotin stabilizers. First, it has a neg-

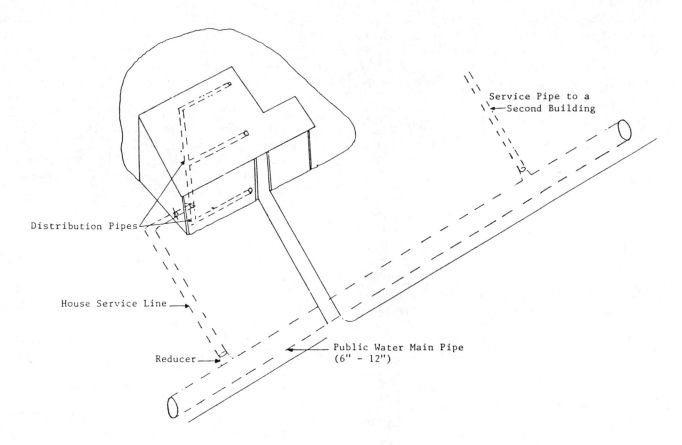

Figure 4–4.   Schematic of a water pipe distribution network.

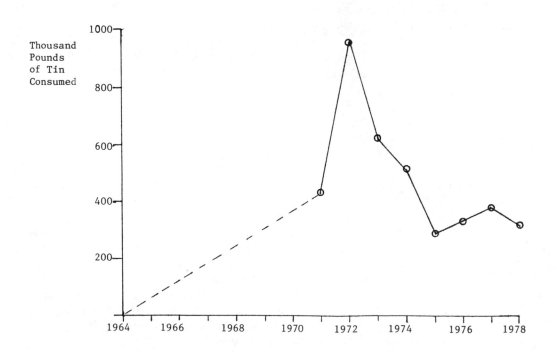

Figure 4–5.   Tin consumption in water pipe, 1964–78. [From table 4–2.]

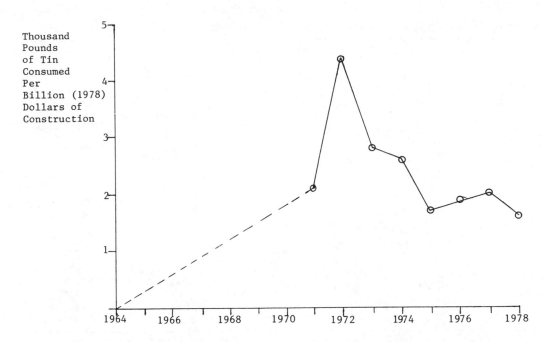

Figure 4–6   Intensity of tin use in water pipe, 1964–78. [Construction expenditures represent private (residential and nonresidential) and public construction in constant 1978 dollars. Residential construction figures were deflated by the GNP deflator for gross private domestic investment for nonfarm structures. Nonresidential and public construction figures were deflated by the GNP deflator for gross private domestic investment for nonresidential structures. From table 4–2; U.S. Council of Economics (1979).]

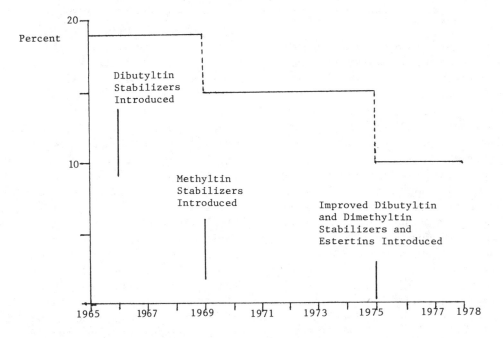

Figure 4–7.   Tin content of organotin stabilizer, 1965–78. [Estimates based on information from industry sources.]

Table 4-3. Apparent Determinants of Tin Consumption in Polyvinyl Chloride Water Pipe, 1965–78

Year	Tin Content of Organotin Stabilizer (percent)	Organotin Stabilizer Content of PVC Water Pipe Compound (percent)	PVC's Share of Plastic Water Pipe (percent)	Plastics' Share of Total Water Pipe Market Times an Unknown Constant (k)[a]	Total Water Pipe Production Divided by an Unknown Constant (k)[b]	Tin Consumption[c] (thousand pounds)
1965	19	2.00	86	100.21	191.42	62.69
1966	19	1.80	86	223.06	191.09	125.37
1967	19	1.60	86	382.99	187.82	188.06
1968	19	1.40	86	549.17	199.59	250.74
1969	15	1.20	86	940.60	215.26	313.43
1970	15	1.00	86	1545.99	188.59	376.11
1971	15	.93	86	1778.72	205.63	438.80
1972	15	.86	93	3637.69	219.70	958.80
1973	15	.80	90	2559.13	223.60	618.00
1974	15	.73	89	2679.82	196.20	512.40
1975	10	.66	91	2856.12	169.00	289.90
1976	10	.60	90	3438.14	179.30	332.50
1977	10	.53	91	4063.34	191.30	374.90
1978	10	.46	91	3883.04	198.90	323.30

Sources: Data on the tin content of organotin stabilizer and on the organotin stabilizer content of PVC water, DWV, and sewer pipe compound are based on information from industry sources. Figures on PVC's share of the various plastic pipe markets for 1971 through 1978 are from the Plastics Pipe Institute (various years). The figures shown for earlier years are estimates based on information from industry sources. Data on the amount of PVC resin used in the manufacture of water, DWV, and sewer pipe since 1971 are also from Plastics Pipe Institute (various years). U.S. construction expenditures in billions of 1978 dollars from U.S. Council of Economic Advisers, *Annual Report of the President, 1978*.

[a]Since information on construction expenditures (Q_t) and plastic pipe production (Y_t) is available, tin consumption in the production of PVC plastic pipe is estimated by modifying the basic identity in the following way:

$$T_t \equiv a_t\, b_t\, c_t\, (k\; d_t)\, (P_t/k)$$
$$\equiv a_t\, b_t\, c_t\, (Y_t/Q_t)\, (Q_t)$$

This column therefore gives the ratio of plastic water pipe production (Y_t), measured in thousands of pounds, to construction expenditures (Q_t), measured in billions of 1978 dollars. For reasons discussed in the text, this ratio is a proxy for plastics' share of the total water pipe market multiplied by an unknown constant (k).

[b]This column shows U.S. construction expenditures (Q_t) in billions of 1978 dollars, which serve as a proxy for total water pipe production divided by an unknown constant (k).

[c]Tin consumption is the product of the tin content of organotin stabilizer, the organotin stabilizer content of PVC pipe compound, and the quantity of PVC compound used in water pipe. The latter is assumed to equal 105 percent of the weight of the PVC resins used in water pipe. Prior to 1971, data on PVC resin consumption in water pipe are not available, and the figures shown for tin consumption are based on the linear trend between 1964 (when tin consumption was negligible) and 1971.

ative effect on tin consumption since lower values of a_t will give lower values of T_t, and a_t has declined over the years. Second, its value does not vary with end use application, so it is the same for PVC water, DWV, and sewer pipe.

Stabilizers are just one of many ingredients that go into a PVC pipe compound and there is little difficulty in switching from one stabilizer to another. No new equipment is required, nor must personnel be retrained. Simply, a slightly different compound mix is extruded. Since new and more efficient stabilizers tend to be adopted quickly once introduced, the stepwise function shown in figure 4-7 probably reflects closely the changes over time in the average tin content of organotin stabilizer consumption.

Technological change was largely responsible for this substantial drop. An appreciation of the important innovations requires some knowledge of organotin chemistry. The general structure of organotins can be represented by:

$$R_2\, Sn\, X_2$$

where R_2 represents an alkyl group, Sn the tin atom, and X_2 an organic group usually bound to tin through an oxygen or sulfur atom. "Butyl" was one of the first alkyl groups used and dibutyltins became known as the first-generation stabilizers. (Table 4-4 presents the salient features of these and other stabilizers.) Other alkyl groups used today include the "octyls" and the "methyls."

Different alkyl groups vary in their ability to combine with the tin atom and lead to compounds with higher or lower tin content based on molecular weight. For instance, the octyltins, because of their particular molecular makeup, are less efficient at stabilization than the butyltins when equal weights of each are used. The methyltins, introduced in 1969, represented a major technological innovation. Their superiority as stabilizers stems from a patented technique of adjusting the tail end of the molecule to produce a "neat" molecule which, though structurally different, (R_2 Sn X_2 no longer applied), has enhanced stabilization properties.

The tail end or "X" of this "neat" molecule represents organic groups, hence the organic nature of these stabilizers. One organic group known as carboxylates does not contain sulfur and is rarely used today for PVC pipe stabilization. A second organic group that does contain sulfur is known

Table 4-4. Major Technological Developments in Organotin Stabilizers

Nomenclature	Generation	Date Introduced	Average Tin Content by Weight (percent)	Properties
Dibutyl-tin	First Generation	1964	18–20	Good lubricating properties.
Octyl-tin	First Generation	1965	18–20	Generally less toxic and less powerful than the butyltins.
Methyl-tin or "Super tins"	Second Generation	1969	15	More efficient than 1st generation types, cheaper to make, worst for toxicity. Superior cost performance than first generation types.
Improved First and Second Generation Stabilizers	Third Generation	1975	10	More efficient than earlier generation types.
Estertins[a]	Third Generation	1975	10	Low toxicity, excellent ultra-violet protection, improved heat distortion, higher impact strength. Very economical.

Source: Communication with industry sources.

[a]"Estertins" is a trademark of AKZO CHEMIE, BV, Netherlands.

as "mercaptides." It is used extensively, since the inclusion of sulfur has a positive effect on stabilization.

Research and development have also focused on changing the structural characteristics of the mercaptides to enhance stability while lowering tin levels. Economization of tin use became increasingly desirable as rising stabilizer demand, which did not put upward pressure on tin prices, was nevertheless accompanied by higher tin prices. The result has been more "third generation" stabilizers. "Estertins" represent one category and their manufacture employs a new production route called "Reverse Ester" technology. This new technology makes "Estertins" more economical to produce, so they are less expensive on the market. It involves synergistic[7] effects with other materials, such as calcium stearate, which enables stabilizers to be used at lower concentrations while embodying smaller amounts of tin (personal interview).

The price of tin has been a major factor stimulating the research and development efforts to reduce the tin content of organotin stabilizers. As figure 4-8 illustrates, it fell at an average annual rate of 10 percent between 1965 and 1968, remained constant between 1969 and 1972, and then rose an average of 14 percent a year between 1973 and 1978.

It is natural to expect rising tin prices to put upward pressure on organotin stabilizer prices and falling tin prices to do the opposite. The magnitude of this pressure depends on the extent to which tin contributes to the cost of these stabilizers. In 1965, when the tin content of organotins was 19 percent, tin was responsible for 18 percent of organotin stabilizer costs. By 1977, tin accounted for 33 percent of organotin stabilizer costs, even though tin content had fallen to 10 percent. While organotin stabilizer prices declined until 1977, as will be shown in the next section, the price of tin rose from 1973 onward. Thus, over most of the 1970s, as figure 4-9 indicates, the cost of tin became an increasingly larger component of the price of organotin stabilizers. This increase provided the incentive for research and development.

Organotin Stabilizer Content of PVC Plastic

There are two ways to influence the organotin content of PVC pipe compound. The processor, having chosen tin stabilizers, may decide for reasons made clear later, to use less tin stabilizer per pound of PVC resin. Alternatively, the processor may elect not to use tin stabilizers at all, opting instead for antimony stabilizers.

Over the past fourteen years, the organotin content of PVC pipe compound has fallen dramatically. Though the data are sketchy, the general trend is illustrated in figure 4-10. In 1965, a typical PVC pipe compound mix contained 2 percent stabilizer by weight. Today that figure is 0.4 percent, or one-fifth the 1965 figure.[8] Thus, determinant b_t, like determinant a_t, has had a negative effect on tin consumption.

---

[7]The phenomenon of synergism occurs when the efficiency of a compound exceeds the sum of the efficiencies of its constituents. Since all the desirable heat and light stability characteristics cannot be found in any one stabilizer and different stabilizers possess complementary properties, combining stabilizers is one way to optimize stabilization. Also, a wide variety of stabilizing actions is possible from different mixtures (Chevassus and DeBroutelles, 1963).

[8]A typical PVC water pipe compound mix today contains 100 parts of resin, 1-5 parts of calcium carbonate, 1 part of calcium stearate, 0.8 part of 165° paraffin wax, 0.15-0.2 part of partially oxidized polyethylene wax, 1 part of titanium dioxide and 0.3-0.4 part of stabilizer.

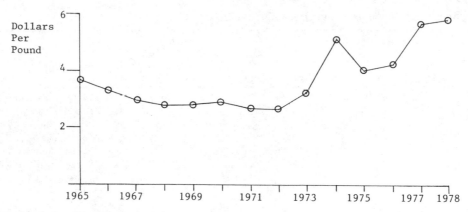

Figure 4–8. New York tin prices in constant 1978 dollars. [From International Tin Council, *Statistical Supplement*, (1969–70); International Tin Council, *Tin Statistics*, (1966–77); and U.S. Department of the Interior, Mineral Commodity Summaries, (1978).]

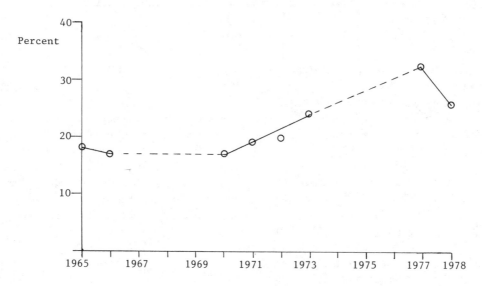

Figure 4–9. Percentage of the total costs of organotin stabilizers represented by tin costs, 1965–78. [From *Modern Plastics* (various years); and information from industry sources.]

There are four principal reasons for the declining organotin use in PVC water pipe compounds. First, the successive generations of stabilizers incorporated technological improvements that enhanced stabilizing efficiency, enabling their use at lower concentrations. For example, the more efficient super-tins reduced stabilizer concentrations 21 percent over conventional types (*Modern Plastics*, September 1971, p. 61). These developments resulted from better understanding of organotin stabilizer chemistry, which was formerly considered a "black art."

Second, advanced extruder technology played a significant role. In the early 1960s, single screw extruders, as discussed earlier, were used to produce plastic pipe, but they subjected the heat-sensitive PVC resin to temperatures exceeding 400°F. When multiple screw extruders were introduced in 1967, maximum processing temperatures were lowered to 350°F. This reduced the requirements for stabilizer protection by as much as 40 percent (*Modern Plastics*, September 1970, p. 98).

Third, PVC compounding was increasingly performed in high intensity mixers, hastening the obsolescence of ribbon blenders. The latter used low-speed spiral blades to mix the ingredients of the pipe compound. Low speeds lessened frictional heat, but had the adverse effect of reducing dispersion of ingredients and mixability. With high intensity mixing, temperature and pressure conditions are such that the resin opens up like popcorn, facilitating the absorption of the compounding ingredients. When mixing is complete,

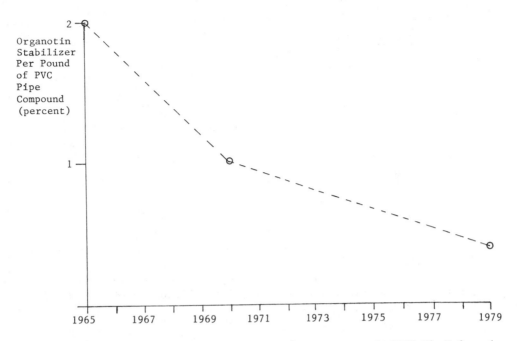

Figure 4–10.   Organotin stabilizer content of PVC pipe compound, 1965–79. (Information from industry sources.)

cooling ensues and each resin particle theoretically comes out as a complete compound. Consequently, each particle is better able to withstand processing temperatures (personal interview).

Fourth, new methods of producing vinyl chloride monomer (VCM), the basic raw material from which PVC is made, have led to better quality resins from which a superior pipe compound can be made. Such process innovations have reduced but not eliminated PVC's susceptibility to heat.

The price of tin has contributed indirectly to the drop over time in the organotin stabilizer content of PVC pipe. Though only 0.4 percent of pipe weight is attributable to stabilizers, the latter's costs contribute 3 percent to pipe costs, of which the price of tin constitutes 1 percent. Pipe manufacturers were still concerned that ever-increasing tin prices would eventually mean higher stabilizer costs for them and a less competitive pipe product, at a time when plastic pipe had to be cheap to encourage its use by skeptics. Therefore, manufacturers supported research and development which led to the four advances in technology described. Again, these developments, which caused b_t to decline, were motivated by the desire to save on stabilizer cost, of which tin is an important part.

Figure 4-11 shows the evolution over the 1965–78 period in the real price of organotin stabilizer, which fell continuously until 1978 despite the rising price of tin after 1972. While the latter exerted upward pressure on stabilizer prices, greater competition in the stabilizer market resulting from the expiration of certain patents helped maintain the downward trend in price.

The higher price of tin naturally encouraged efforts to minimize and even eliminate its use in stabilizers. The synergistic effects mentioned earlier in connection with "reverse ester" stabilizers are produced by replacing tin with cheaper metals, such as cadmium, barium, and antimony. Many of these stabilizers have recently been approved for usage in PVC water pipe and are likely to replace the traditional organotins (personal interview).

In this regard, another important factor has influenced the amount of organotin stabilizer used per pound of PVC pipe compound, preventing this apparent determinant of tin consumption from declining even more than shown in figure 4-10. Before stabilizers can be used in water pipe, they must have National Sanitation Foundation (NSF) approval with regard to their safety. Health concerns center around the elution of toxic substances in potable water in particular, but also in any waste that people may later be exposed to. Consequently, maximum limits of inorganic chemicals in potable water have been set. Until 1979, NSF approved only organotins for use in potable water pipe. Lead stabilizers, capable of performing the same functions and at half the price, were barred from water pipe on grounds that lead might migrate into the water. However, lead stabilizers are used extensively in Europe for plastic water pipe.

The exclusion of lead may have been, at least in part, due to pressure from domestic producers. Stabilizer technology in the United States evolved around organotins, and there was a clear desire by certain domestic interests to minimize the use of lead stabilizers, whose technology was perfected in Europe. These interests may have exaggerated

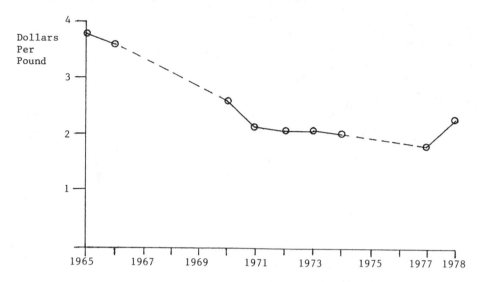

Figure 4–11.   **Organotin stabilizer prices in constant 1978 dollars.** [From *Modern Plastics* (various issues).]

lead's toxicity problem to encourage its exclusion from water pipe, the largest sector of the PVC pipe market.

The exclusion of lead stabilizers secured for the "tins" a virtual monopoly in PVC water pipe until 1978. Antimony stabilizers appeared as potential substitutes for tin stabilizers since 1975. On a weight basis they are not as efficient as tin stabilizers, requiring 12 to 16 percent more antimony for equivalent performance. Yet the significantly lower cost of antimony results in stabilizers superior to organotins from a cost-efficiency standpoint. For the next several years, little antimony was used because consumer loyalty to the "tins" and skepticism of antimony inhibited the latter's adoption. After years of testing, antimony finally gained approval in water pipe in 1979, the year after the data analyzed in this study stop.

Although lower prices encouraged the use of antimony stabilizers, it was the changing attitudes of pipe producers, who grew more receptive as their knowledge and experience with antimony accumulated, that was mainly responsible for its acceptance. The use of antimony stabilizers is expected to continue to grow in the future, further reducing the average content of organotin stabilizers in PVC pipe.

PVC's Share of the Plastic Water Pipe Market

PVC is by far the leading plastic material used today for transporting drinking water. Its market share in this application, illustrated in figure 4-12, has always been greater than 85 percent. PE, chlorinated polyvinyl chloride (CPVC), and polybutylene (PB) are some of the other plastics used for transporting potable water, but their market shares have been significantly lower. Because PVC gained market share in this application, determinant c_t has had a positive effect

on tin consumption, since the more PVC that is used, *ceteris paribus*, the greater the demand for (tin) stabilizers.

The 90 percent share of the plastic water pipe market that PVC has consistently maintained over the 1970–78 period has not been greatly influenced by the price of tin. The latter simply accounts for too small a fraction, less than 1 percent, of the final cost of PVC pipe. Moreover, the battle for the plastic water pipe market has historically been between PVC and PE. The prices of these resins, as figure 4-13 shows, have over time moved closely together.

PVC pipe dominates the plastic water pipe market for other reasons. As figure 4-14 shows, PVC and CPVC have much greater tensile and compressive strength than their competitors in the water pipe market. Since water pipe, particularly water mains, requires the material withstand high pressures, PVC is naturally preferred and this largely explains its 90 percent market share of plastic water pipe.

Other properties help to establish PVC's superiority mechanically. Elongation (in response to temperature change) measures the percentage increase in length per unit length. As illustrated in appendix 4-1, PVC and CPVC have the lowest elongation of the materials involved. Flexural strength measures the resistance of a plastic strip to breaking as it is bent across its axis. Since impact resistance generally increases with flexural strength, PVC enjoys superiority in both property classifications. Overall, the mechanical properties of PVC make it better able to withstand continuous deformation than its competitors in the plastic water pipe market.

PVC gained advantage in water pipe because of its thermal properties. Knowledge of these properties is important since plastics are subjected to much heat during processing and sometimes ultraviolet radiation afterward. Selection of

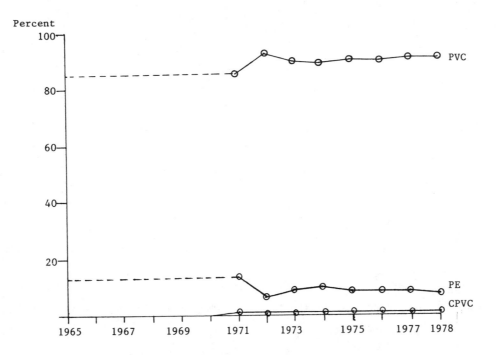

Figure 4–12.   Market share of various plastic materials used in plastic water pipe, 1965–78. [Polybutylene's market share has been less than 1 percent since it was introduced in 1975. From Plastics Pipe Institute (various years).]

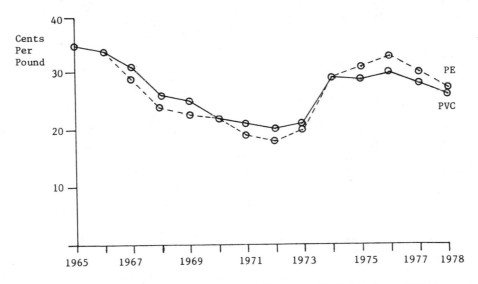

Figure 4–13.   PVC and PE prices in constant 1978 cents, 1965–78. [From U.S. International Trade Commission, *Synthetic Organic Chemicals, U.S. Production and Sales* (various years).]

a plastic for hot water transport, for instance, is also based in part on thermal properties.

Thermal conductivity measures the rate of heat transfer through a plastic body. Naturally, the lower the value, the lower the conductance. PVC and especially CPVC, as figure 4-15 indicates, have relatively low thermal conductivities. This is both a blessing and a curse. It explains, for example, why CPVC is chosen for hot water transport. It also explains

why early attempts to process PVC posed enormous difficulties. Besides the degradation brought about by exposure to heat in the extruder, its poor heat conduction makes it difficult to obtain a uniform product (Mason and Manning, 1945).

Other interesting thermal properties include specific heat, which is the heat required to raise the temperature of 1 gram of the plastic 1 degree, and the coefficient of thermal ex-

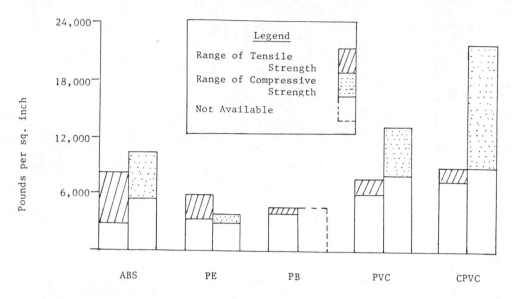

Figure 4–14. Range of compressive and tensile strengths for various plastic pipe materials.

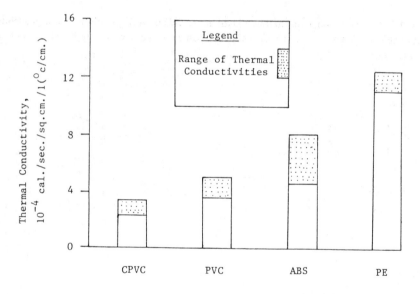

Figure 4–15. Range of thermal conductivities for various plastic pipe materials.

pansion, which measures the change in unit dimension caused by a 1: degree temperature rise. These properties are listed in the appendix. PVC's lower specific heat reduces extruder heating costs while its lower thermal expansion makes it more stable *in situ*.

A most important thermal property is flammability, which measures burning rate. The importance of this cannot be overstressed since some plastics are notorious for quick burning. In fact, concern over flammability of plastics is so extensive that building codes stress fire protection. Regu-

lations require that the plastics used in homes have a certain resistance to burning. Mason and Manning (1945) point out that plastics which contain a high proportion of chlorine in the molecule show reduced flammability. Rigid PVC has a high chlorine content, making it inherently fire resistant. With PE, for instance, flame retardants (chlorine-containing compounds, hydrated inorganic salt or antimony oxide) must be added. It is principally this requirement for self-extinguishability that makes PVC the natural choice for plastic pipe in buildings.

Plastics' Share of the Water Pipe Market

Since data on the total tonnage of water pipe manufactured annually in the United States are not available, construction expenditures are used as a proxy on the assumption that year-to-year fluctuations in the production of water pipe closely follow changes in real construction expenditures. The amount of plastic and copper water pipe produced annually per billion dollars of real construction expenditures is shown in figure 4-16. Plastic pipe, though primarily PVC, also includes PE, CPVC, and a small amount of polybutylene (PB). The last two are used mainly for transporting hot water. Unfortunately, information on other materials used to transport drinking water, such as prestressed concrete, asbestos cement, ductile iron and steel, is not available, and so cannot be included in figure 4-16.

The trend in plastic pipe consumption is consistently upward save for drops in 1973 and 1978. This increasing use of plastics per billion dollars of construction is strong evidence that plastics' share of the water pipe market has been growing at the expense of other competing materials. Additional evidence for this conclusion comes from a survey of water main pipe installations conducted by the American City Magazine (1975). Its results show that plastic pipe accounted for 9.3 percent of all the new water main pipe installed during 1974. In contrast, its share of the pipe already in place was only 1.1 percent.

The upward growth shown in figure 4-16 between 1964 and 1978 represents an important period in the history of plastic water pipe development. Prior to 1964, the calcium-zinc compounds used as stabilizers proved inadequate, and plastic water pipe then had a very short life span. It was the introduction of organotins in 1964 that gave plastic pipe producers a durable product and triggered the subsequent rise in consumption. This growth was further stimulated by the appearance of the superior methyltins in 1969 and the widespread use of multiscrew extruders by 1972. Plastics, then, continued to gain market share throughout the 1970s as the initial skepticism of many potential users was overcome.

One material to suffer at the hands of plastics was copper. Figure 4-16 shows that its intensity of use per billion dollars of construction has declined or remained stagnant since 1965.

A number of factors have influenced the speed with which plastic has penetrated the water pipe market:

1. One major consideration is price. Although data do not exist to compare all types of pipes on a cost per foot basis, the available information does indicate that plastic pipe enjoys a cost and price advantage in many applications. This conclusion is confirmed by industry sources.

Figure 4-17 shows that the price per pound of copper has consistently been more than double that of PVC. Behind this price difference is the complex, costly, capital-intensive process for producing copper, compared with the relatively simple process for making PVC pipe compound. In addition, its extrusion process is a simple one-step operation compared with a three-step operation for copper. Thus, fabrication costs are significantly higher for copper. As a result, copper pipe may cost more than three times as much as plastics in equivalent applications.

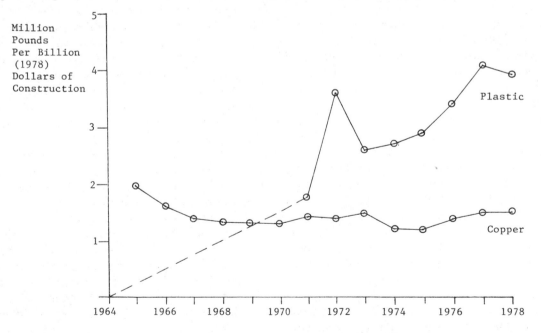

Figure 4–16. Plastic and copper water pipe consumption per billion (1978) dollars of construction expenditures. [From Plastics Pipe Institute (various years); Copper Development Association (1979); and U.S. Council of Economics (1979).]

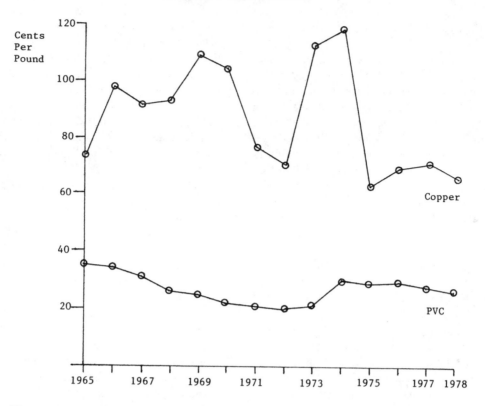

Figure 4–17.   Prices for copper and PVC plastic, 1965–78. [The price shown for copper is the U.S. producers' price. From *Engineering and Mining Journal* (various issues); U.S. International Trade Commission, *Synthetic Organic Chemicals, U.S. Production and Sales* (various years).]

Enhancing plastics' cost-competitiveness in recent years are escalating energy costs, since metals are more energy intensive than plastics. This is true even after taking into account the energy conservation resulting from the recycling of metals (DuPont, 1979). Though no section of the economy is spared from rising energy costs, plastics' lesser dependence improves its cost competitiveness.

2. Technological change has also enhanced the attractiveness of plastic pipe by making it stronger, expanding its range of applications, and by reducing its production costs. For instance, advances in polymer chemistry have produced resins capable of withstanding high temperatures. Chlorinated polyvinyl chloride (CPVC) introduced in 1970 and polybutylene (PB) introduced in 1975 are two developments that have allowed plastic pipe to transport hot water. Production costs have been reduced by improved resins, advances in extruder technology, and numerous other innovations (personal interview).

Technological developments, of course, have not been confined to plastics. The threat of plastics stimulated research and development aimed at reducing manufacturing costs for other kinds of pipe. In copper, for example, there has been a trend to thinner wall copper tube. In 1965, a 2-inch diameter copper tube 2.125 feet long could be extruded

from 1 pound of copper. By 1978, a length of 2.70 feet could be obtained (Copper Development Association, n.d.).

3. Plastics possess a combination of physical and chemical properties that produce cost savings when used as water pipe. The durability of plastics compared with metals and other materials stems, not from raw physical strength as measured by crushing and piercing tests, but rather from their nonelectrolytic action and their rust and corrosion resistance.

Of particular importance for water transport is the stain-resistance of plastics and the fact that water does not wet plastic surfaces. Combined with smoothness, this increases flowability, making it possible for smaller diameter plastic pipe to transport water at rates that ordinarily would require larger diameter metal pipe.

Low density is another physical property that favors plastics. Light weight makes installation easy, reducing manpower requirements. Replacement costs are lower for the same reason. Finally, the flexibility of PE and PB permits the use of coiled lengths, eliminating fittings in many instances.

4. An important factor that has frequently delayed and inhibited the use of plastic pipe is approval by code authorities. Many townships, counties, and cities have their

own regulations pertaining to building practices. Others adopt model codes which are written by private groups. Among the latter are the Basic Building Code (BBC) maintained by building officials and by the Code Administrators International based in Chicago, the National Building Code (NBC) of the American Insurance Association, the Standard Building Code (SBC) of the Southern Building Code Congress based in Birmingham, and the Uniform Building Code (UBC) of the International Conference of Building Officials. At the federal level, the Department of Housing and Urban Development (HUD) and the Federal Housing Administration (FHA) are, by their ability to influence legislation, the major bodies influencing construction standards.

Code approval for plastic pipe was not readily forthcoming, and this delayed its adoption. For instance, the UBC still does not accept plastic pipe for hot water transport, and other codes approve only under certain conditions (*Modern Plastics*, April 1977, p. 47). Also, the use of plastic pipe in the municipal water main market was prevented for years because the American Water Works Association (AWWA) did not approve its use until 1975. Additional evidence comes from a U.S. National Bureau of Standards (1978, p. 19) investigation. It reports:

In the 1960s thermoplastic piping for residential plumbing systems lacked widespread acceptance by American code authorities because metallic piping was already proven, acceptable and available, whereas thermoplastic piping was unproven in this application and many designers and installers lacked the knowledge, experience and initiative to utilize it properly. But gradually a body of supporting data has been accumulated, so that in the past few years the material has been increasingly accepted for various applications.

The increasing acceptance of plastics reflects changing attitudes by writers of model codes. Since many jurisdictions accept the model codes, opposition to the use of plastics can have widespread effects. However, there is little consistency in the model codes. For example, the UPC does not approve ABS for potable water service pipe, while the other model codes accept it, but for cold water only. While some codes take definite positions on some plastics, others for various reasons, do not. And those who approve certain materials impose numerous limitations relating to height, location, type of waste, type of occupancy, fire rating of building, pipe wall thickness, special installation rules, special test requirements, combustibility, etc. (National Bureau of Standards, Building Science Series 111, 1978). As a result of this confusion, many local jurisdictions that cannot afford to carry out tests to assess the adequacy of plastics stick to their old codes and materials, restricting plastics' use.

5. Union opposition has also retarded the growth of plastic pipe. The reduced manpower requirements associated with the use of plastic pipe has worried union leaders and bitter court battles have been waged to restrict plastics' use.

6. Contractor inertia stemming from ignorance has been another problem. Many considered plastics a fire hazard. Some even thought plastics were susceptible to rodent attacks. Such reservations, combined with pressure from vested interests, delayed the acceptance of plastics by building code bodies and local independent building groups. The diverse standards that developed reflect this confusion. They also made it difficult for suppliers to standardize their products, because adjacent localities often imposed different standards.

7. Population density is another variable influencing the use of plastic pipe. In large cities and densely populated areas, water utilities resort to extremely large diameter pipe made of prestressed concrete to meet strength requirements and realize economies of scale. Plastics, given the current state of technology, cannot usually compete in large diameter (12 inches and more) markets. However, *some* municipalities have rapidly changing populations and have opted for 12-inch diameter pipes made of plastic. This has two advantages: (1) it reduces the initial investment, and (2) plastic pipe can easily be replaced if necessary.

Total Pipe Production

Since data on the total tonnage of pipe manufactured in the United States are not available, construction expenditures are used as a proxy on the assumption that year-to-year fluctuations in the production of pipe closely follow changes in real construction expenditures.

Over the 1965–78 period, the trend in construction expenditures has been cyclical. Hence the impact on tin consumption of apparent determinant P_t, which is proxied by Q_t, has not been consistent. Sometimes it has been positive; other times negative. Overall, its impact has been modestly positive.

Since construction expenditures are used as a proxy for total water pipe production, the factors that affect the latter are assumed to be identical with those factors that affect construction expenditures. Many of these variables are macroeconomic in nature but others are demographic.

Briefly, the demand for water (DWV and sewer) pipe depends on the demand for homes and commercial buildings, which in turn depends on the general economic conditions in the economy, including interest rate levels. High interest rates make borrowing more expensive, reducing the demand for housing and pipe. During a boom period, construction expenditures rise, causing the demand for pipe to increase.

The important demographic characteristics are population size, density, and age distribution. Total pipe demand can be expected to be positively correlated with population size, a high average age, and low population density. The latter results in the extension of pipe lines to remote areas where single-family homes are prevalent.

Comparative Effects of Apparent Determinants

Tin consumption in the production of water pipe has increased from minute amounts in 1964, used primarily for experimental purposes, to 323,000 pounds in 1978. As pointed out earlier, this increase reflects the combined effects of five apparent determinants.

Three of these determinants have increased tin consumption in water pipe: the growth in the water pipe market (P_t), the increase in plastics' share of the water pipe market (d_t), and the rise in PVC's share of the plastic water pipe market (c_t). Together these apparent determinants would have increased tin consumption to 2.7 million pounds in 1978, rather than the actual 323,000 pounds, had the negative effects of the other two apparent determinants not offset part of their impact.

To assess their relative importance, the separate effect of each apparent determinant on tin consumption over the 1964–

78 period is estimated. For the "positive determinants," this is done by calculating the change in tin consumption each alone would have caused had all the other apparent determinants remained unchanged at their 1964 values. For the "negative determinants," the separate effect is calculated by the increase in tin consumption each would have caused had it alone remained at its 1964 value, assuming all other determinants changed to their 1978 value.

Moreover, when two or more apparent determinants increase tin consumption, there is a multiplicative effect that must be added to their separate effects. For example, if one determinant increased in value by 10 percent and another by 10 percent, then their combined effect is not 20 percent, but 21 percent. The additional 1 percent is the multiplicative effect. Similarly, when two or more apparent determinants reduce tin consumption, there is a multiplicative effect, which is negative rather than positive.

These calculations are shown in appendix 4-3, and illustrated in figure 4-18. The amount of tin used in water pipe in 1964 was zero because in that year plastic had not yet significantly penetrated the water pipe market. Determinant

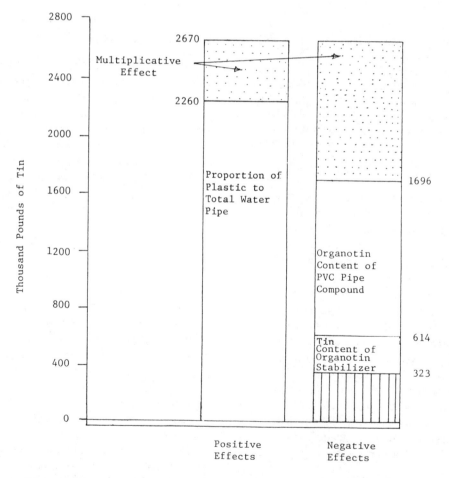

Figure 4–18. Effects of changes in apparent determinants on tin consumption in water pipe, 1964–78.

d_t, plastics' share of water pipe, was zero, and hence tin consumption was also zero.

The first column of figure 4-18 shows the amount of tin— 2.7 million pounds—that would have been required by the producers of (PVC) water pipe in 1978 had only those apparent determinants changed whose influence increased the use of tin. The separate effect of one of these apparent determinants is also shown in this column: the growth of plastics in water pipe by itself increased tin consumption by 2.3 million pounds. In fact, without a change in this apparent determinant, none of the changes in the other determinants would have increased tin use. Hence the increases in c_t, PVC's share of plastic pipe, and P_t, total pipe production, helped tin consumption in water pipe only because d_t, plastics' share of pipe, changed from zero. Consequently, the separate effects of c_t and P_t cannot be shown in figure 4-18 since their individual contributions to tin consumption in water pipe cannot be separated. The sum total of their contribution is labeled in figure 4-18 as the multiplicative effect.

The second column indicates the actual amount of tin— 323,000 pounds—consumed in water pipe in 1978, along with the effects of those apparent determinants reducing the use of tin. The two negative apparent determinants are the tin content of organotin stabilizers, a_t, and the organotin stabilizer content of PVC plastic, b_t. The separate effect of a_t was a negative 291,000 pounds in 1978, much less than the 1.1 million pound reduction caused by b_t. The multiplicative effect between a_t and b_t caused a further reduction in tin consumption of 974,000 pounds.

In terms of percent change, the growth of plastics in water pipe had the greatest single effect on tin consumption in water pipe. This is clearly illustrated in figure 4-18. The negative effect of the organotin concentration in PVC compound was second in importance.

Among the more important underlying factors responsible for the above changes in apparent determinants is technological change. Indeed, it was the introduction of organotin stabilizers in 1964 that made it possible for plastic to penetrate the water pipe market. Subsequent improvements in tin stabilizers and the introduction of the multi-screw extruders, by lowering production costs, further enhanced the competitiveness of plastic pipe. They also reduced the requirements for tin in stabilizers and for stabilizers in plastic pipe compounds.

The price of tin has motivated research and development aimed at reducing the tin content of stabilizers. This is understandable since tin has accounted for between 18 and 35 percent of stabilizer costs. Beyond this, tin costs are of minor importance, contributing less than 1 percent to the final costs of pipe.

The price of plastic, on the other hand, has influenced the share of plastic pipe in the water pipe market. Relatively low compared to the price of copper, steel, and the other

materials used to make pipe, it has been a major factor encouraging builders and construction contractors to switch to plastic pipe.

The superior properties of PVC have allowed it to dominate the plastic pipe market. While PE is at times used in small-diameter, low-pressure, flexible applications because of its softness, PVC is stronger and the obvious choice for the pressure applications common for water pipe. In addition, PVC's self-extinguishability gives it a major advantage.

Among the available stabilizers, those containing tin have benefitted from health considerations. Indeed, had it not been for the concern over the migration of lead into drinking water, little or no tin stabilizers would be in use, for lead stabilizers can perform the same functions at substantially lower costs.

The most important underlying factor retarding the use of tin in water pipe is the slow rate of code approval. Though its effect on tin consumption cannot be precisely quantified, it has been substantial. Opposition from unions and others with interests in maintaining the traditional sectors of the water pipe market, along with uncertainty and ignorance over the long-run durability and safety of plastic pipe, have also kept the use of plastic pipe from expanding even faster than it has.

Drain-Waste-Vent Pipe

The drain-waste-vent (DWV) market for plastic pipe provides many interesting contrasts with the market for water pipe.[9] Particularly significant differences are found in the speed of code approval, the types of competing materials, the growth in market share of plastic pipe, and the competition between tin and other (antimony) stabilizers.

Estimated tin consumption in DWV pipe is shown in figure 4-19 for the 1971–78 period. The dashed line from 1964, when no tin was consumed in this end use, to 1971, covers the years for which reliable PVC consumption data are unavailable. This figure indicates a cyclical pattern of tin usage with peaks in 1974 and 1977. Figure 4-20 shows that the intensity of tin use, measured by the pounds of tin consumed in DWV pipe per billion dollars of real construction expenditures, has followed a similar pattern.

These changes in tin consumption reflect the combined effect of changes over time in the tin content of stabilizers (a_t), the organotin stabilizer content of PVC pipe compound (b_t), PVC's share of the plastic DWV pipe market (c_t), plastics' share of the total DWV pipe market (d_t), and the

[9]A drain is a pipe installed outside of a building that carries liquid waste, while a waste pipe refers to an inside installation conveying waste that is free of fecal matter. Vents are installed inside but protrude outside. They facilitate the efficient functioning of the drainage system by providing a circulatory stream of air that avoids back pressures.

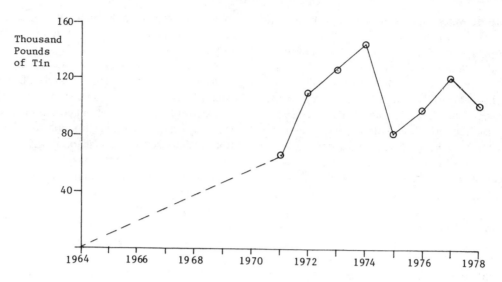

Figure 4–19.   Tin consumption in drain-waste-vent pipe, 1964–78. [From table 4–4.]

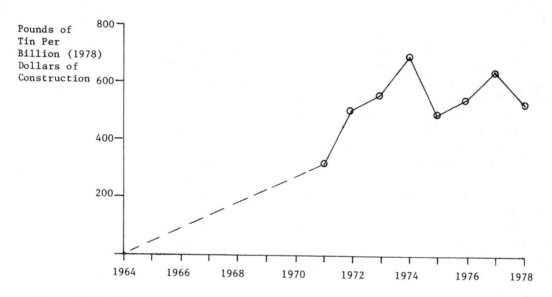

Figure 4–20.   Intensity of tin use in drain-waste-vent pipe, 1964–78. [From table 4–4 and U.S. Council of Economics (1979).]

size of the DWV pipe market (P_t) as approximated by real construction expenditures. Table 4-5 illustrates the changes in these apparent determinants over time. Again, since the actual size of DWV pipe production is unknown, the influence of changes in the last two apparent determinants— plastics' share of the total DWV pipe market and the size of the DWV market—are approximated by the ratio of plastic DWV pipe production to construction expenditures (Y_t/Q_t) and by construction expenditures (Q_t).

Since real construction expenditures were examined in the last section, the changes over time in this apparent determinant are not considered again here. Similarly, the tin content of the organotin stabilizer is the same whether the

product being manufactured is water pipe or DWV pipe, and so the trends in this determinant described earlier apply as well to the analysis here of DWV pipe.

It is the other three apparent determinants, whose influence is not the same for DWV and water pipe, that are responsible for the interesting differences in tin usage patterns between these two types of pipe. These are examined in the following sections.

Organotin Stabilizer Content of PVC Plastic

The changes over time in this determinant, which are illustrated in figure 4-10 for water pipe, are the same for

Table 4-5. Apparent Determinants of Tin Consumption in Polyvinyl Chloride DWV Pipe, 1965–1978

Year	Tin Content of Organotin Stabilizer (percent)	Organotin Stabilizer Content of PVC DWV Pipe Compound (percent)	PVC's Share of Plastic DWV Pipe (percent)	Plastics' Share of Total DWV Pipe Market Times an Unknown Constant (k)[a]	Total DWV Pipe Production Divided by an Unknown Constant (k)[b]	Tin Consumption[c] (thousand pounds)
1965	19	2.00	10	129.50	191.42	9.42
1966	19	1.80	20	144.14	191.09	18.84
1967	19	1.60	25	197.98	187.82	28.26
1968	19	1.40	30	236.51	199.59	37.67
1969	15	1.20	35	347.24	215.26	47.09
1970	15	1.00	40	499.41	188.59	56.51
1971	15	.93	38	612.90	205.63	65.93
1972	15	.86	28	1413.38	219.70	112.16
1973	15	.80	42	1122.43	223.60	126.19
1974	15	.73	56	1148.12	196.20	137.39
1975	10	.66	58	1266.09	169.00	82.19
1976	10	.60	54	1699.23	179.30	97.33
1977	10	.53	60	2024.05	191.30	123.13
1978	10	.44	68	1746.60	198.90	103.33

Sources: Same as table 4-3.

[a]Since information on construction expenditures (Q_t) and plastic pipe production (Y_t) is available, tin consumption in the production of PVC plastic pipe is estimated by modifying the basic identity in the following way:

$$T_t \equiv a_t\, b_t\, c_t\, (k\, d_t)\, (P_t/k)$$
$$\equiv a_t\, b_t\, c_t\, (Y_t/Q_t)\, (Q_t)$$

This column therefore gives the ratio of plastic DWV pipe production (Y_t), measured in thousands of pounds, to construction expenditures (Q_t), measured in billions of 1978 dollars. For reasons discussed in the text, this ratio is a proxy for plastics' share of the total DWV pipe market multiplied by an unknown constant (k).

[b]This column shows U.S. construction expenditures (Q_t) in billions of 1978 dollars, which serve as a proxy for total DWV pipe production divided by an unknown constant (k).

[c]Tin consumption is the product of the tin content of organotin stabilizer, the organotin stabilizer content of PVC pipe compound, and the quantity of PVC compound used in DWV pipe. The latter is assumed to equal 105 percent of the weight of the PVC resins used in DWV pipe. Prior to 1971, data on PVC resin consumption in DWV pipe are not available, and the figures shown for tin consumption are based on the linear trend between 1964 (when tin consumption was neglibile) and 1971.

DWV pipe as for water pipe with the sole but important exception of 1978. In that year, antimony stabilizers were first used in significant quantities in DWV pipe.

In 1975, antimony mercaptides or stabilizers appeared as potential substitutes for tin stabilizers. On a weight basis they are not as efficient as tin stabilizers, requiring 12 to 16 percent more antimony for equivalent performance. Yet the significantly lower cost of antimony results in stabilizers superior to organotins from a cost-efficiency standpoint. For the next several years, little antimony was used because consumer loyalty to the "tins" and skepticism of antimony inhibited the latter's adoption. In 1978, however, antimony stabilizers captured nearly 4 percent of the market for PVC-DWV pipe. As a result, the organotin content of PVC compound for DWV pipe averaged 0.44 percent in 1978 compared with the 0.46 percent shown in figure 4-10 for water pipe.

The penetration of antimony stabilizers into the DWV market was in large part due to the fact that in 1978 they sold for $1.40 per pound compared to $2.25 for tin stabilizers. Also, the attitudes of pipe producers grew more receptive as their knowledge and experience with antimony accumulated. The use of antimony stabilizers is expected to continue to grow in the future, further reducing the average content of organotin stabilizers in PVC-DWV pipe.

PVC's Share of the Plastic DWV Pipe Market

Consumption in the DWV plastic pipe market of PVC and ABS, the two plastics competing for this market, is illustrated in figure 4-21 for the years 1967–78. Early in this period, ABS enjoyed the larger market share. In 1967, for example, 75 percent of the total plastic used for DWV pipe was ABS. Over time, however, the use of PVC has grown more rapidly. As a result, its share of the market surpassed that of ABS in 1974, and was more than twice as large by 1978.

A number of underlying factors have contributed to the rising market share of PVC plastic:

1. The relative prices of PVC and ABS resins have all along favored PVC. As figure 4-22 indicates, the price of ABS has typically been double that of PVC on a per pound basis. There are two principal reasons for this. One is the disparity in raw material prices, illustrated in figure 4-23. ABS contains acrylonitrile, butadiene, and styrene, all of which are more expensive on a per pound basis than vinyl

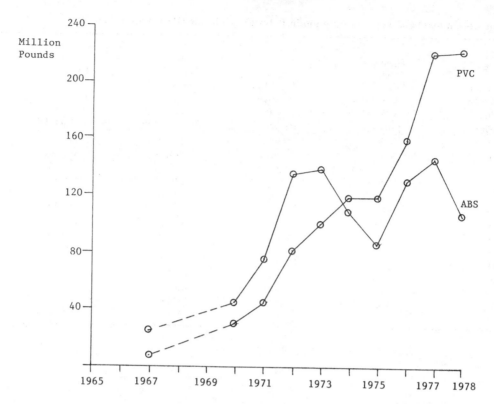

Figure 4–21.   PVC and ABS plastic consumption in the drain-waste-vent pipe market, 1965–78. [From Plastics Pipe Institute (various years); *Modern Plastics* (various issues).]

chloride, the raw material from which PVC is made. Second, the two-step polymerization process required for ABS resins makes them more expensive than those plastic resins, such as PVC, that use a one-step process.

2. Despite the lower price of PVC, the physical properties of ABS initially offset this cost disadvantage and led to its widespread use in the DWV plastic pipe market. While processors shied away from PVC due to its poor heat resistance, ABS had a reputation for being tough, with high strength and temperature resistance. These are excellent machining qualities, and ease of fabricating ABS reduced the price disadvantage of its resins. New market entrants were willing to experiment with it first before risking the relatively delicate PVC process.

The glaring discrepancy in price between PVC and ABS resins, however, stimulated research to improve PVC's extruder performance. While technological changes that ensued had no effect on resin price, they did influence product price. Briefly, these changes involved sophisticated extruder designs and advanced stabilizer chemistry. The net result was a superior PVC pipe compound that was exposed to less heat during production and able to withstand more heat.

3. Code approval also influenced PVC's market penetration. PVC initially gained approval more slowly than ABS for two reasons. First, ABS had a proven record in the mobile home industry where it had been used extensively

in DWV applications since the early 1960s. PVC, therefore, had to overcome the loyalty and consumer inertia built up by ABS's earlier acceptance. Second, PVC's processing difficulties, already described, were an obstacle.

The problems PVC had in gaining code acceptance can be easily illustrated. The International Association of Plumbing and Mechanical Officials (IAPMO) which co-operates with the Uniform Building Code (UBC) approved ABS-DWV in the early 1960s but not PVC until October 1968 (*Modern Plastics*, January 1969, p. 50). By 1967, ABS-DWV had 145 approvals from national, regional, and local bodies compared to 32 for PVC (*Modern Plastics*, October 1967, p. 104–105). Over time the situation for PVC improved. By 1969, there were 13 state and 300 local code approvals for PVC-DWV pipe. These numbers increased to 18 and 500 respectively by 1970 and to 40 and 900 by 1971 (*Modern Plastics*, February 1972, p. 42). This increased acceptance of PVC enabled it to compete in more markets with ABS, and was a crucial factor in its growing market share.

4. Shortages of acrylonitrile, butadiene, and styrene, the three raw material ingredients of ABS, have also aided PVC. Styrene, which comprises 50-75 percent of ABS, comes from benzene. William Storck (1979) writing in *Chemical and Engineering News* notes: "The benzene situation is still one of the biggest worries for ABS producers. Users of benzene and its derivatives in the chemical industry have

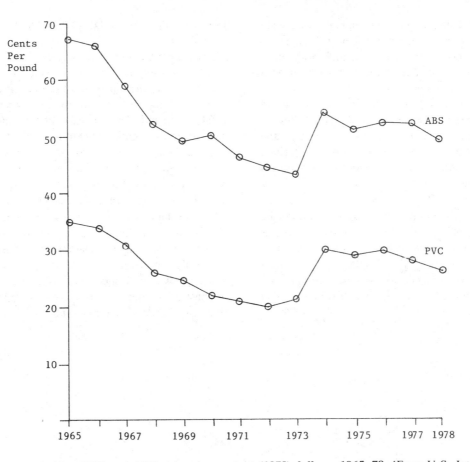

Figure 4–22.   PVC and ABS prices in constant (1978) dollars, 1965–78. [From U.S. International Trade Commission, *Synthetic Organic Chemicals, U.S. Production and Sales* (various years).]

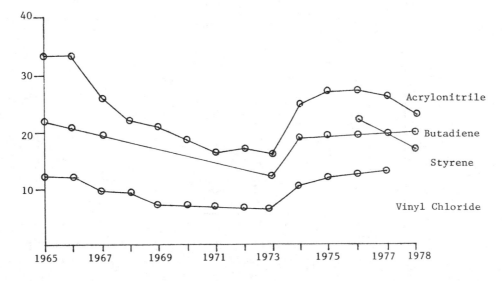

Figure 4–23.   Raw material prices of ABS and PVC resins in constant cents (1978). [From U.S. International Trade Commission, *Synthetic Organic Chemicals, U.S. Production and Sales,* (various years).]

102 MATERIAL SUBSTITUTION

had to compete hard with the gasoline makers, who want the compounds to raise the octane number of unleaded gasoline.''Storck also points out that low feedstock availability has on a number of occasions during the 1970s forced ABS producers to operate below capacity.

Butadiene has also had supply problems. Its availability for ABS production has at times been adversely affected by fluctuating imports, strong demand from tire makers, and the use of ethylene feedstock to produce lighter materials. Since ethylene and butadiene are coproducts, the latter results in lower butadiene production (*Chemical and Engineering News*, July 30, 1979, p. 8).

The above difficulties, coupled with tight acrylonitrile supplies in 1974 and 1975, led to a shortage of ABS resins and pipe. This caused an irretrievable loss of market share to PVC-DWV pipe suppliers. Many ABS pipe producers became pessimistic and switched to PVC pipe production, a change that cost considerations would probably have brought about at some later date.

Plastics' Share of the DWV Pipe Market

Among the many materials used for DWV pipe are plastics, CISP, steel, copper, and reinforced concrete. Figure 4-24 shows the production of plastic and CISP DWV pipe per billion dollars of real construction expenditures over the 1964–78 period. The omission of the other types of DWV pipes from the figure is due to the lack of data. It is known, however, that very little copper and steel are still used in the DWV market. Large diameter reinforced concrete pipes or culverts are used for highway and airport drainage to the tune of 2.6 billion pounds a year, or 13 million pounds per billion dollars of construction expenditures. The focus on plastic and cast iron soil pipe, while necessitated by data considerations, may be desirable since they tend to compete for similar applications.

The growth in plastic pipe consumption per billion dollars of construction depicted in figure 4-24 suggests that plastics' share of the DWV pipe market has been increasing over the

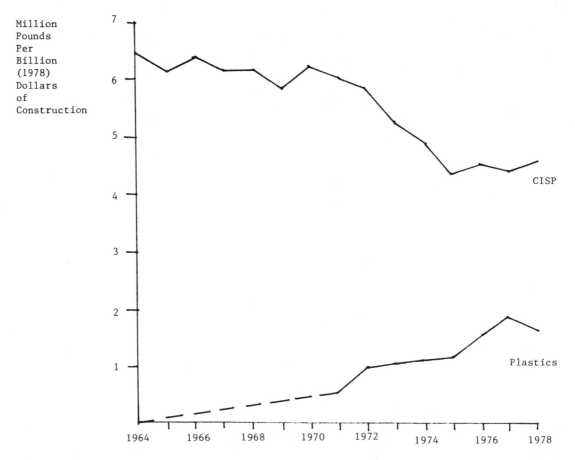

Figure 4–24. Plastic and cast iron soil pipe production for the DWV market per billion (1978) dollars of construction expenditures, 1964–78. [From Plastics Pipe Institute (various years); Cast Iron Soil Pipe Institute, *Annual Sales Statistics* (various years); U.S. Council of Economics (1979).]

last decade. Part of this increase apparently has come at the expense of cast iron soil pipe, as its share of the DWV market seems to have declined since 1970.

Three underlying factors have been particularly important in affecting the growing share of plastic pipe in the DWV market:

1. The relatively low price of plastic compared to alternative materials has already been examined. Industry sources confirmed that the cheapness of plastic pipe in terms of dollar cost per foot has been a major factor stimulating its continued growth.

2. Plastic pipe offers many of the properties desired for DWV pipe, such as corrosion resistance, smoothness to facilitate flow, and easy joining. Of particular importance is plastics' lightness. A 20-foot length of 2-inch plastic pipe, for example, weighs only 15 pounds. A similar pipe made of copper is 21 pounds, steel 73 pounds, and cast iron 80 pounds (Copper Development Association, 1979, p. 17). The importance of weight for installation costs has already been stressed.

Plastics' lightness, however, is a mixed blessing. It has long been known that sound transmission is related to material density. Lightweight plastics readily experience molecular vibration, which allows sound to travel easily through and along the pipe walls. Dense materials like cast iron transmit sound less readily because they are harder to excite. The end-to-end cementing of plastic pipe also facilitates sound transmission. Cast iron pipes are joined such that the pipe ends are not in contact. This discontinuity at the joint adds to the quietness of CISP-DWV systems. A study by Polysonics (1970) on "Noise and Vibration Characteristics of Soil Pipe Systems" concludes that vibration drops ranging from 11 to 20 decibels (dB) per joint are realized with cast iron pipe while "lightweight soil pipe systems such as copper (with sweat joints) and plastic pipe (with glued joints) give essentially no drop in vibration or noise across joints." Thus, CISP is often selected over plastic for DWV systems where minimal plumbing noise is desired.

3. Much has already been said about the refusal by many code bodies to accept plastics in water pipe. In the DWV market, plastics have suffered the same fate. Table 4-6 shows the results of a 1979 survey by *Plastics Focus* magazine. It is clear that CISP is still the preferred material, particularly in apartment and commercial buildings as well as below ground. The higher the building and the more people that use it, the greater the opposition seems to be to plastics. Refusal to use plastics below ground stems from fears that the pipe may be crushed. Whatever the reason for the opposition, it is clear that this limited approval of plastics has slowed its penetration of the DWV pipe market.

Comparative Effects of Apparent Determinants
Small quantities of tin were first used in stabilizers for the DWV market in 1964. Usage grew until 1974, dropped

Table 4-6. Code Acceptance of CISP, PVC and ABS for DWV Applications, 1979
(Percentage of jurisdictions approving)

Application	Cast Iron Soil Pipe[a]			PVC	ABS
	Type I	Type II	Type III		
Single-family homes	96	75	93	93	88
Apartments	95	75	91		
Low-rise apartments				75	71
High-rise apartments				56	57
Commercial buildings	93	75	91	68	61
Below ground	73	87	87	b	b

Source: Plastics Focus, 1980.
[a]Types I, II and III refer to the different ways cast iron soil pipe and fittings can be connected. Type I is known as the NO-HUB joint. It is a new plumbing concept that involves the use of a one-piece neoprene gasket, a stainless steel shield, and retaining clamps. Type II is known as the compression joint. It uses hub and spigot pipe and a one-piece rubber gasket. Type III is known as the lead and oakum joint. This joint uses molten lead and oakum (Cast Iron Soil Pipe Institute, 1976).
[b]Not available but known to be less than figures shown for cast iron soil pipe.

substantially in 1975, partially recovered in 1976 and 1977, and then dropped again in 1978. Despite these recent fluctuations, over the 1965–78 period, tin consumption in this end use increased substantially.

The relative effects of the five apparent determinants responsible for this upward trend in tin use are shown in figure 4-25. The growth in the DWV pipe market, PVC's share of the plastic DWV pipe market, and plastics' share of the total DWV pipe market are the three determinants responsible for the increased usage of tin. The separate effect of the last of these three determinants is 116,000 pounds of tin. The multiplicative effect of all three is also shown in figure 4-25.

Reductions in both the tin content of stabilizers and the stabilizer content of PVC pipe compound have had a negative influence on tin consumption, reducing it to 103,000 pounds, substantially below what it otherwise would have been. Of the two, the decline of the stabilizer content of PVC compound had the greater impact.

The important underlying factors responsible for these changes in apparent determinants are for the most part the same as those affecting tin consumption in the water pipe market. However, two important differences should be noted.

First, PVC did not dominate plastic pipe production for the DWV market from the beginning, as it did in the water pipe market, but rather over time increased its market share at the expense of ABS plastic. This change, which enhanced the use of tin in this application, came about primarily as a result of new technology that improved the physical properties of PVC and made it more competitive with ABS pipe.

Second, the penetration of antimony stabilizers accentuated the decline in the organotin content of PVC in the DWV market during 1978, and is likely to continue to do so in the future. While the effects of this development may eventually affect the organotin content of PVC used for

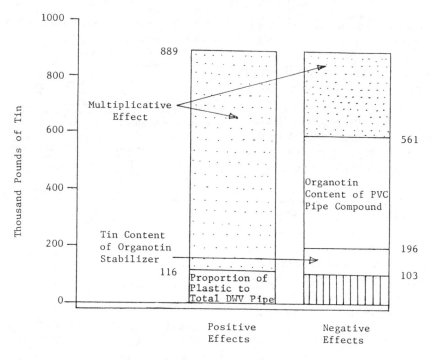

Figure 4–25. Effects of changes in apparent determinants on tin consumption in drain-waste-vent pipe, 1964–78.

plastic water pipe as well, code approval and acceptance are likely to come more slowly in that sector of the pipe market, helping to maintain and prolong the use of tin.

Sewer Pipe

Sixteen percent of plastic pipe production in 1978 went into sewer applications, compared to 19 percent for DWV applications and 40 percent for water applications. The relatively low figure for plastic sewer pipe reflects its slow growth during the 1960s. Since 1971, however, the production of plastic sewer pipe has increased at a rate several times that of plastic DWV and water pipe.

Figure 4-26 shows that the consumption of tin in sewer pipe has followed the sharp upward trend of plastic sewer pipe production during the 1970s. Over the years 1971–78, the growth of tin averaged 29 percent in sewer applications, compared to 4.5 percent in both water and DWV applications. The intensity of tin use in sewer pipe, measured in figure 4-27 by the pounds of tin used per billion dollars of real construction expenditures, has also risen rapidly since 1971.

Two of the five apparent determinants responsible for these changes in tin consumption vary over time in the same manner for the sewer market as for the water and DWV markets: namely, the tin content of the organotin stabilizer (a_t); and total sewer pipe production (P_t), whose fluctuations are approximated by real construction expenditures. In ad-

dition, the organotin content of PVC (b_t) has, as was the case for DWV pipe, followed the same pattern shown in figure 4-10 for water pipe, except in 1978 when antimony stabilizers captured an estimated 2 percent of the market which organotins had previously monopolized. Since the changes in these determinants over time, shown in table 4-7, have already been examined along with the underlying factors responsible, this section focuses on the two remaining apparent determinants.

PVC's Share of the Plastic Sewer Pipe Market

PVC faces competition from polyethylene (PE) and styrene rubber (SRP) in the plastic sewer pipe market. It held only 16 percent of this market in 1971, but managed to increase this figure to 50 percent by 1978. In the meantime, styrene rubber, the dominant plastic in 1971, was nearly forced out of the market. These changes are illustrated in figure 4-28.

The demise of styrene rubber was primarily a matter of cost. Though actual cost data are not available, the many ingredients used in the multistaged production process for styrene rubber pipe make it more than twice as expensive as PVC or PE pipe. In the 1960s, the use of stryene rubber was favored by code-making authorities, largely because there was little experience with PVC and PE to evaluate. As these newer plastics proved themselves in other markets, code approval for sewage applications followed.

PE was able to match PVC's rapid growth during the 1970s because of similarly low costs and a number of de-

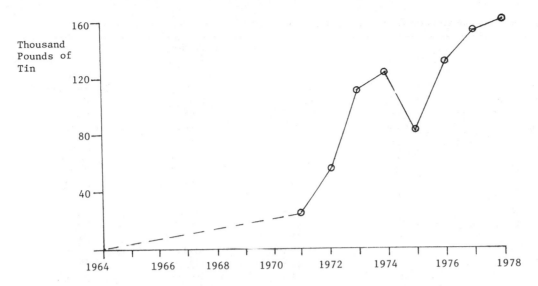

Figure 4–26.   Tin consumption in sewer pipe, 1964–78. [From table 4–6.]

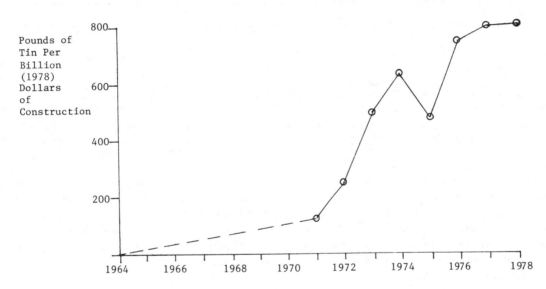

Figure 4–27.   Intensity of tin use in sewer pipe, 1964–78. [From table 4–6, U.S. Council of Economics (1979).]

sirable attributes. In particular, the flexibility of PE and its easy butt-joining techniques reduce installation costs. In addition, its use was stimulated by European-based technology that permits the manufacture of large diameter PE pipe.

Plastics' Share of the Sewer Pipe Market

Many materials are used for sewer pipes. The more prominent ones include cast iron soil pipe, asbestos cement, vitrified clay, reinforced concrete, and plastics. Unfortunately, data are not available for vitrified clay, reinforced concrete, and asbestos cement. However, data do exist for

plastic and cast iron soil pipe, and figure 4-29 illustrates their production per billion dollars of construction expenditures since 1964. The substantial rise in plastic sewer pipe per billion dollars of construction since 1971 suggests that plastics' share of the sewer market has been increasing. In contrast, the use of cast iron soil pipe per billion dollars of construction remained stable until 1972 and then declined. These trends suggest that plastics are replacing CISP and other types of materials in the sewer pipe market.

Many of the underlying factors influencing the penetration of plastics in this market have been discussed in earlier sections on DWV and water pipe. The relatively low price of plastic pipe, for example, again has encouraged its use

Table 4-7. Apparent Determinants of Tin Consumption in Polyvinyl Chloride Sewer Pipe, 1965–78

Year	Tin Content of Organotin Stabilizer (percent)	Organotin Stabilizer Content of PVC Sewer Pipe Compound (percent)	PVC's Share of Plastic Sewer Pipe (percent)	Plastics' Share of Total Sewer Pipe Market Times an Unknown Constant (k)[a]	Total Sewer Pipe Production Divided by an Unknown Constant (k)[b]	Tin Consumption[c] (thousand pounds)
1965	19	2.00	18.8	26.54	191.42	3.63
1966	19	1.80	18.8	59.09	191.09	7.26
1967	19	1.60	17.4	109.61	187.82	10.89
1968	19	1.40	16.7	163.66	199.59	14.51
1969	15	1.20	17.1	273.78	215.26	18.14
1970	15	1.00	17.4	442.28	188.59	21.77
1971	15	.93	16.1	549.98	205.63	25.40
1972	15	.86	22.1	922.66	219.70	57.79
1973	15	.80	39.3	1040.50	223.60	109.72
1974	15	.73	41.5	1383.27	196.20	123.33
1975	10	.66	45.5	1586.58	169.00	80.52
1976	10	.60	52.1	2351.68	179.30	131.81
1977	10	.53	47.8	3134.70	191.30	151.92
1978	10	.45	49.7	3574.71	198.90	159.30

Sources: Same as table 4-3.

[a]Since information on construction expenditures (Q_t) and plastic pipe production (Y_t) is available, tin consuption in the production of PVC plastic pipe is estimated by modifying the basic identity in the following way:

$$T_t \equiv a_t\, b_t\, c_t\, (k\, d_t)\, (P_t/k)$$
$$\equiv a_t\, b_t\, c_t\, (Y_t/Q_t)\, (Q_t)$$

This column therefore gives the ratio of plastic sewer pipe production (Y_t), measured in thousands of pounds, to construction expenditures (Q_t), measured in billions of 1978 dollars. For reasons discussed in the text, this ratio is a proxy for plastics' share of the total sewer pipe market multiplied by an unknown constant (k).

[b]This column shows U.S. construction expenditures (Q_t) in billions of 1978 dollars, which serve as a proxy for total sewer pipe production divided by an unknown constant (k).

[c]Tin consumption is the product of the tin content of organotin stabilizer, the organotin stabilizer content of PVC pipe compound, and the quantity of PVC compound used in sewer pipe. The latter is assumed to equal 105 percent of the weight of the PVC resins used in sewer pipe. Prior to 1971, data on PVC resin consumption in sewer pipe are not available, and the figures shown for tin consumption are based on the linear trend between 1964 (when tin consumption was negligible) and 1971.

in sewer applications. Similarly, code approval and union opposition have inhibited its acceptance.

The importance of concrete and clay in the sewer market, and the rapid rise of plastics, are both in large part due to the low costs of these materials. Stronger and more expensive materials, such as CISP, are usually reserved for high-pressure sewer main applications. However, sewer systems are designed where possible to take advantage of gravity flow. Since 90 percent of the sewer systems today employ gravity flow, this severely restricts the use of CISP and other high-pressure materials.

Sewage characteristics also influence the choice of material. High temperatures, low velocities, and the consequent formation of sulfuric acid caused by hydrogen sulfide oxidation create an atmosphere that readily dissolves iron, concrete, and asbestos cement (Steel and McGhee, 1979). Plastics and vitrified clay have the advantage of being corrosion resistant, but they suffer from diameter limitations imposed by technology. Clay pipe has a 42-inch diameter maximum and plastics a 12-24 inch maximum. Concrete has no diameter constraints but it must be lined when used in an acidic environment.

One factor that has kept plastic sewer pipe from growing even faster is the lack of evidence on effective life. Sewer

pipes made from clay are known to have an average useful life of 100 years. Plastics were introduced only in the mid-1960s, and have no long-term track record. So many builders believe the use of plastics may involve some risk, for broken sewer lines can cause extensive damage and create serious legal problems.

Comparative Effects of Apparent Determinants

Between 1964 and 1978, the consumption of tin in plastic sewer pipe rose from zero to about 160,000 pounds. Figure 4-30 shows the individual and multiplicative effects of the five apparent determinants. The increase in total sewer pipe production (P_t), the share of PVC in the plastic sewer pipe market (c_t), and the share of plastic pipe in the total sewer market (d_t) are the three determinants responsible for this rise in tin usage. Particularly important is the increase in plastics' share of the total sewer market. Had it remained at zero, none of the changes in the other apparent determinants could have increased tin use. Figure 4-30 also indicates that tin consumption in 1978 was substantially less than it would have been if the tin content of organotin stabilizers (a_t) and the organotin stabilizer content of PVC pipe compound (b_t) had not fallen. Of the two, the latter

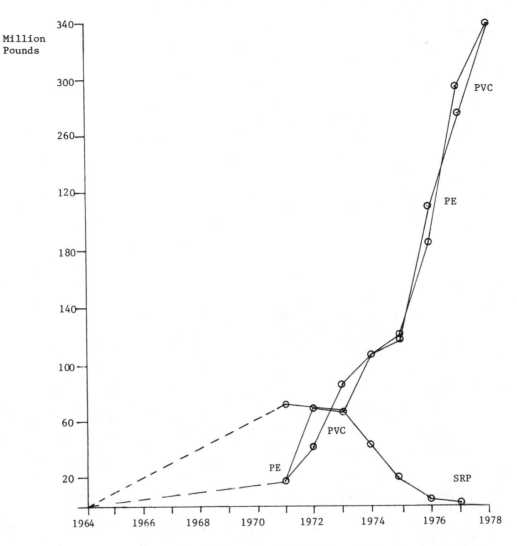

Figure 4–28. Plastic consumption in sewer pipe, 1964–78. [From Plastics Pipe Institute (various years).]

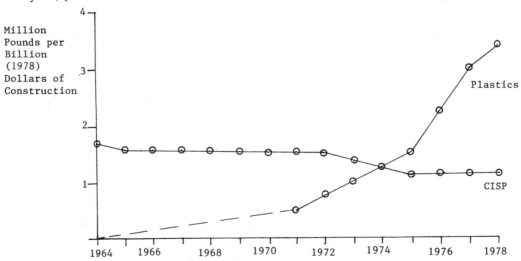

Figure 4–29. Plastic and cast iron soil pipe production for the sewer market per billion (1978) dollars of construction expenditures. [Same as figure 4–24.]

107

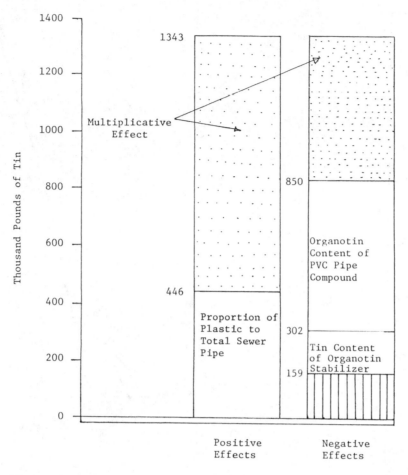

Figure 4–30. Effects of changes in apparent determinants on tin consumption in sewer pipe, 1964-78

was again the more important determinant reducing tin consumption.

The important underlying factors causing these changes have all been discussed in connection with water and DWV pipe. They include new technological developments that have improved the properties of plastic sewer pipe and reduced its costs, the inherent cheapness of the principal raw material (vinyl chloride) used to produce PVC, the recent introduction of antimony stabilizers, and the difficulties in gaining approval for plastic pipe from code authorities, contractors, and unions.

Tin Consumption in Stabilizers for the Pipe Industry

Seventy-five percent of the plastic pipe produced in 1978 was for water, DWV, and sewer applications. Together these three sectors of the pipe industry consumed 1.3 billion pounds of PVC plastic, containing 586,000 pounds of tin in organotin stabilizers. This is a new use for tin. In the

early 1960s, plastic pipe had not penetrated the markets examined in this chapter.

Three apparent determinants are responsible for this new use of tin. First, plastic pipe entered the water, DWV, and sewer sectors and increased its share of these markets over the intervening period. Second, PVC increased its share of the plastic pipe market for DWV and sewer applications, and continued to dominate water applications. Third, construction expenditures in the United States grew, increasing the country's need for pipe. While these determinants stimulated the consumption of tin in plastic pipe production, their positive influence was offset to a considerable extent by a reduction over time in the amount of tin used in organotin stabilizers and in the amount of organotin stabilizer used in PVC plastic.

The underlying factors causing the apparent determinants and thus ultimately the consumption of tin to change have been many and varied. Obviously, the rising real price of tin in recent years has not prevented the demand for tin in this new area from growing rapidly. This is not really surprising. Since tin is used in such small quantities, its price contributes less than 1 percent to the final cost of PVC pipe. Moreover, vinyl chloride, the major raw material used to produce PVC, is very inexpensive. As a consequence, PVC pipe is relatively cheap compared with other plastics and to other materials used to make pipe. In addition, PVC's mechanical, thermal, and chemical properties enhance its attractiveness.

The price of tin is much more important at the stabilizer level, where it has accounted for as much as a third of stabilizer costs. Here it has been an important factor motivating new tin-saving innovations, including the development of second and third generation organotin stabilizers, as well as the recent tin-free antimony stabilizers that may completely eliminate tin from the plastic pipe market within the next few years.

These and other technological changes appear to be the most important underlying factors affecting tin usage in the plastic pipe market. The rapid growth of plastic pipe in these markets was originally made possible by the development of organotin stabilizers, and then further stimulated by subsequent advances that improved the quality of PVC pipe and reduced its costs.

Finally, a host of institutional factors have influenced tin consumption. Particularly important in this regard are the delays and barriers imposed by building code requirements. The hundreds of authorities responsible for setting standards, by their fragmented nature, ensure that approval to use new materials will occur slowly and in a piecemeal fashion. Many local code authorities are inadequately financed and understaffed. Not having the resources to test new materials adequately, they tend to stick with traditional materials.

In a few instances, the lethargy and conservatism introduced by code requirements have worked to the advantage of tin. For example, lead stabilizers, though considerably cheaper, have been enjoined from the water pipe market on the grounds that lead might migrate into the water and pose a health hazard. More recently, code approval has delayed the substitution of antimony for tin stabilizers. Generally though, code requirements have delayed and prevented the use of plastic pipe, and in the process dampened the use of tin in this end use. Further reducing tin use were two factors: opposition from unions and others with interests in maintaining the traditional sectors of the water pipe market, and uncertainty and ignorance over the long-run durability and safety of plastic pipe.

Appendixes to Chapter 4

Appendix 4-1. Mechanical and Thermal Properties of Plastics Used in Pipe

Properties	PVC	CPVC	ABS	PE	PB
			Mechanical		
Tensile Strength, p.s.i.	6,000–7,500	7,500–9,000	2,500–8,000	3,100–5,500	3,800–4,400
Comprehensive Strength, p.s.i.	8,000–13,000	9,000–22,000	5,200–10,200	2,700–3,600	—
Elongation, percent	40–80	4.5–65.0	20–100	20–1,300	0.26–0.50
Flexural Yield Strength, p.s.i.	10,000–16,000	14,500–17,000	4,000–14,000	—	—
			Thermal		
Thermal Conductivity, 10^{-4} cal./sec./sq.cm./ 1 (°C/cm.)	3.5–5.0	3.3	4.5–8.0	11.0–12.4	—
Specific Heat, cal./°C/gm.	0.25–0.35	0.33	0.3–0.4	0.55	0.45
Thermal Expansion, 10^{-5} in./in./°C	5.0–10.0	6.8–7.6	6.0–13.0	11.0–13.0	15.0
Flammability, Burning rate, in./min.	—	—	0.6–1.0	1.00–1.04	1.08

Source: Modern Plastics Encyclopedia, 1975–76.

Appendix 4-2. Additives Used in Plastic Pipe Production

Antioxidants—These are low cost, nontoxic chemicals that are added in small concentrations to polymers to mitigate the deleterious effects of embrittlement caused by aging, and to minimize oxidative and photo-initiated degradation. Such degradation is possible during processing, fabrication, storage, and use, and may even result from outdoor exposure. It is evidenced by color changes and alteration in physical and chemical properties. Antioxidants must have good heat and light stability and be compatible with the polymer to prevent easy extraction.

Colorants—These compounds affect opacity, transparency, and the overall attractiveness of the final product. They are usually inert to preserve their own properties and those of the other components of the mixture. Among the factors influencing colorant performance are: resin type, other ingredients present and their concentration, chemical resistance, type and temperature of processing. Colorants must be durable and resistant to various light intensities.

Flame Retardants—These are usually compounds of antimony, bromine, chlorine, nitrogen, phosphorus or boron which are added to a pipe mix to increase burning resistance so that fire safety regulations can be met. Flame retardants must be added to ABS and PE polymers but not rigid PVC, which has a good fire resistance because of its high chlorine content.

Fillers—These are added to resins to improve properties. Fillers can reduce moisture absorption, and increase mechanical strength, heat resistance and electrical characteristics. When resins are in short supply, fillers are used as extenders. Many types of fillers are used. Among them are glass, carbon, cellulose fillers, other carbohydrates, metallic oxides, silicates and various inorganic compounds.

Lubricants—These compounds determine ease of processing by affecting fusion speed, viscosity, and friction between the polymer and machine parts. They improve resins chemically by enhancing stability, and fabricated products are physically improved by increasing impact strength. Lubricants are categorized as internal or external. External lubricants lower the coefficient of friction between polymer and metal and thus function as a lubricant. Internal lubricants help the polymer to flow by reducing viscosity. Internal friction generates great heat, causing the compound to undergo a more severe heat cycle.

Stabilizers—These compounds are added to PVC resins prior to fabrication to help them withstand processing heat since high temperatures may cause decomposition and loss of properties. Many different kinds of stabilizers can be used, including metallic soaps, various lead and antimony compounds and organotin chemicals. The last group dominates the PVC pipe market for reasons explained in the text. Stabilizers are categorized as heat or light stabilizers. The latter function is often performed by antioxidants.

Source: Modern Plastics Encyclopedia, 1975–76, pp. 162–215.

Appendix 4-3

The derivation of the numbers for figure 4-12 is provided in this appendix. The methodology for calculating the numbers for DWV and sewer pipe is the same, and so these calculations are not presented.

The basic identity is:

$$T_t \equiv a_t \, b_t \, c_t \, d_t \, P_t \qquad (1)$$

where the variables a_t . . . . P_t are defined as in section 4. Now let

$$P_t = kQ_t \qquad (2)$$

where Q_t = construction expenditures in billions of 1978 dollars in year t.

k = an unknown constant.

Also let

$$d_t = Y_t/(kQ_t) \qquad (3)$$

where Y_t = the amount of plastic water pipe produced in year t, measured in thousands of pounds.

From (1), (2), and (3),

$$T_t \equiv a_t \, b_t \, c_t \, (Y_t/k \, Q_t) \, (kQ_t) \qquad (5)$$
$$\equiv a_t \, b_t \, c_t \, Y_t$$

From (3) and (5),

$$T_t \equiv a_t \, b_t \, c_t \, d_t \, (k \, Q_t) \qquad (6)$$

Define T^* as the amount of tin that (PVC) water pipe producers would have required in 1978 had only d_t changed. Since tin use in plastic water pipe in 1964 was zero, T^* measures the separate effect of d_t. Then

$$T^* = a_{64} \, b_{64} \, c_{64} \, d_{78} \, (k \, Q_{64}) \qquad (7)$$

Now

$$Y_{78} = d_{78} \, (k \, Q_{78})$$

from equation (3)

Therefore

$$d_{78} \cdot k = \frac{Y_{78}}{Q_{78}} = \frac{769136842}{198.9} \qquad (8)$$

Substituting (8) and (7) and using the values of a_t, b_t, c_t, and Q_t for 1964 gives,

$$T^* = 0.19 \times 0.02 \times 0.86 \times \frac{769136842}{198.9} \times 178.87$$
$$= 2.3 \text{ million pounds.}[1]$$

[1]Numbers are rounded here for convenience.

Now, equation (5) may be rewritten as follows:

$$T_t \equiv a_t \, b_t \, Z_t \qquad (9)$$

where Z_t equals $c_t Y_t$, or the amount of PVC pipe compound used in water pipe in year t. Setting a_t and b_t equal to their values in 1964 and Z_t to its value in 1978 gives,

$$T_{78} = a_{64} \, b_{64} \, Z_{78}$$

where T_{78} is the amount of tin (PVC) water pipe producers would have required in 1978 had only those apparent determinants that positively affected tin consumption changed. It therefore includes the separate effect of d_t as well as the (positive) multiplicative effect explained in the text.

$$T_{78} = 0.19 \times 0.02 \times 702736842$$
$$= 2.7 \text{ million pounds.}$$

Effects of Negative Determinants

Actual tin consumption in 1978 was 323,000 pounds. From equation (9),

$$T_{78} = a_{78} \, b_{78} \, Z_{78}$$
$$= 0.10 \times \frac{0.46}{100} \times 702736842$$
$$= 323,000 \text{ pounds.}$$

If a_t had remained at its 1968 value while all other apparent determinants changed, tin consumption in 1978 would have been,

$$T^*_{78} = a_{64} \, b_{78} \, Z_{78}$$
$$= 0.19 \times \frac{0.46}{100} \times 702736842$$
$$= 614,000 \text{ pounds.}$$

The separate effect of a_t is defined as the amount by which tin consumption in 1978 would have increased had a_t remained unchanged from 1964 assuming all other determinants changed as they did. Thus, the separate effect for a_t is:

$$T^*_{78} - T_{78} = 614 - 323 = 291,000 \text{ pounds.}$$

The separate effect of b_t is similarly calculated. First, calculate tin consumption assuming all the determinants changed but b_t,

$$T^{**}_{78} = a_{78} \, b_{64} \, Z_{78}$$
$$= 0.10 \times 0.02 \times 702736842$$
$$= 1.4 \text{ million pounds.}$$

The separate effect of b_t then equals:

$$T_{78}^{**} - T_{78} = 1405 - 323 = 1.1 \text{ million pounds.}$$

Finally, the sum of (1) actual tin consumption in 1978, (2) the separate effect of a_t, (3) the separate effect of b_t, and (4) the multiplicative effect of a_t and b_t must equal the amount of tin (2.7 million pounds) that would have been consumed in 1978 had only the positive apparent determinants changed. Consequently, the multiplicative effect between a_t and b_t equals $2670 - 323 - 1082$ or 974,000 pounds.

References

American Bureau of Metal Statistics, annual, *Non-Ferrous Metal Data*, New York.

American Can Company, 1953, "First Quarterly Financial Report 1953."

American City Magazine, 1975, *Survey of Water Main Pipe in U.S. Utilities Over 2,500 Population*, Pittsfield, Mass.: Morgan-Grampian Publishing.

American Concrete Pipe Association, 1951, *Concrete Pipe Handbook*, Chicago, Illinois.

American Metal Market, annual, *Metal Statistics*, New York.

Army Industrial College, 1945, *Tin*, (August), Washington, D.C.

Baker, Howard L., 1979, Market Analyst, National Steel Corporation. Personal interview.

Beverage World, monthly, East Stroudsburg, Pa.: Keller Publishing.

Binner, Burton C., Analyst, PA International Management Consultants. Personal interview.

Bruekel, L. P., 1979, Vice President, Tin Mill Products, National Steel Corporation. Personal interview.

Busch Center, 1976, *A Study of the Impacts on the United States of a Ban on One-Way Beverage Containers*, prepared for the U.S. Brewers Association, vol. I and II, Philadelphia: University of Pennsylvania.

Callahan, J., 1965, "New Way to Coat Steel," *Modern Packaging*, vol. 38, no. 5, p. 158.

Casey, William J., Industrial Manager, Vinyl Additives, M & T Chemicals. Personal interview.

Cast Iron Soil Pipe Institute, 1967, *Cast Iron Soil Pipe & Fittings Handbook*, vol. I, Washington, D.C.

———, annual, *Annual Sales Statistics*, McLean, Virginia.

Chemical Week, 1956, "Vying for Solder's Jobs," vol. 97, no. 21, pp. 42–44.

———, 1957, "Organo-tins: The Little Slice with Big Hopes." vol. 7, no. 1, p. 51.

Chemical and Engineering News, July 30, 1978, p. 8.

Chevassus, Fernand, and Roger DeBroutelles, 1963, *The Stabilization of Polyvinyl Chloride*. London: Edward Arnold, Ltd., 385 pp.

Church, F., 1960, "Man of the Year, William K. Coors," *Modern Metals*, vol. 15, no. 12, pp. 88–98.

———, 1972, "Coors Goes All-Aluminum," *Modern Metals*, vol. 28, no. 12, pp. 64–82.

———, 1972, "The Metal Can Market," *Modern Metals*, vol. 23, no. 10, pp. 61–65.

Chynoweth, A. G., 1976, "Electronic Materials: Functional Substitutions," *Science*, vol. 191, no. 4228, pp. 725–732.

Clay Sewer Pipe Association, 1946, *Clay Pipe Engineering Manual*, Columbus, Ohio.

Code of Federal Regulations, 1971, Title 29: Labor, 1910–1960, Washington, D.C.: U.S. Government Printing Office.

Copper Development Association, 1979, *Annual Data 1979*, New York.

———, 1979, *Copper Tube for Plumbing, Heating, Air Conditioning, and Refrigeration*, New York.

———, no date, *New Methods to Reduce Cost of Copper and Brass Automotive Radiators*, data sheet 804/7, New York.

Corplan Associates, 1966, *A Study of the Soft Drink Industry, 1965–1970*, American Bottlers of Carbonated Beverages, Washington, D.C.

Demler, F. R., 1980, "The Nature of Tin Substitution in the Beverage Container Industries," Ph.D. Dissertation, The Pennsylvania State University, University Park.

Dennis, R., 1979, Vice President and General Manager, Beverage Packaging, American Can Company. Personal interview.

DuPont Company, 1979, "Plastics' Use More Than Tripling," unpublished paper, Public Affairs Department, Delaware.

Eckes, A. E., Jr., 1979, *The United States and the Global Struggle for Minerals*, Austin: University of Texas Press.

Engineering and Mining Journal, monthly, New York: McGraw-Hill.

Environmental Protection Agency, 1972, *The Beverage Container Problem, Analysis and Recommendations*, Office of Research and Monitoring, Washington, D.C.

Ewart, W. F., 1978, "Steel versus Aluminum in the Container Market," paper presented to members of the Committee on Promotion and Market Development, International Iron and Steel Institute, Brussels.

Federal Register, 1975, "Lead Occupational Exposure: Proposed Standard," vol. 40, no. 93, Washington, D.C.: U.S. Government Printing Office, pp. 45934–45948.

———, 1978, "Final Standard for Occupational Exposure to Lead," vol. 43, no. 220, Washington, D.C.: U.S. Government Printing Office, pp. 52952–53014.

———, 1979, "Lead in Food: Advance Notice of Proposed Rulemaking," vol. 44, no. 171, Washington, D.C.: U.S. Government Printing Office, pp. 51233–51242.

Forbes, M. K., 1965, "The Manufacture of Vehicle and Industrial Radiators: Part III—Manufacturing Methods," *Tin and Its Uses*, no. 69, pp. 11–15.

Fisher, F. M., P. H. Cootner, and M. N. Baily, 1972, "An Econometric Model of the World Copper Industry," *Bell Journal of Economics*, vol. 3, pp. 568–609.

Gillett, H. W., 1940, "Tinplate and Solder," *Metals and Alloys*, vol. 12, no. 5, pp. 641–645.

Glass Packaging Institute, 1978, *Annual Report*, Washington, D.C.

Goeller, H. E., and A. M. Weinberg, 1976, "The Age of Substitutability," *Science*, vol. 191, no. 4228, pp. 683–689.

Gregory, Marcus, 1979, Commercial Research, United States Steel Corporation. Personal interview.

Hartwell, R. R., 1952, "Trends in the Use of Tin in the Container Industry," *Symposium on Tin*, Special Publication No. 141, American Society for Testing Materials, Philadelphia, pp. 57–61.

Hiers, G. O., 1931, "Soft Solders and Their Applications," *Metals and Alloys*, vol. 2, no. 5, pp. 257–261.

Hoare, W., E. Hedges, and B. Barry, 1965, *The Technology of Tinplate*, New York: St. Martins Press.

Hotchner, S. J., and G. G. Kamm, 1967, "High Tin Fillet Cans for Improved Product and Container Quality," *Food Technology*, vol. 21, no. 6, pp. 95–100.

Houssner, C. E., and E. T. Johnson, 1946, "Substitutes for Tin in the Manufacture of Automobiles," *Automotive and Aviation Industries*, vol. 94, no. 6, pp. 158+.

Howell, W. A., 1979, Vice President of Tin Mill Products, United States Steel Corporation. Personal interview.

International Tin Council, 1971, *Statistical Supplement*, London.

———, 1978, *Tin Statistics, 1967–1977*, London.

———, forthcoming, *Tin Consumption in the United Kingdom, Part II, Solder*, London.

Ireland, J., 1943, "The Future of Tin in Solder," *Tin and Its Uses*, no. 14, p. 10.

Johnsen, M. A., 1962, "The Technical Aspects of Steel and Tinplate Aerosol Containers: Parts I–IV," *Aerosol Age*, June-September, reprint.

———, 1979, Vice President of R & D, Peterson/Puritan, Inc. Personal interview.

———, unpublished, *Aerosol Packaging Materials and Container Technology*.

Katz, Philip C., 1978, "The State of the Industry," *The Brewer's Digest*, no. 10, pp. 44–48.

———, 1979, Vice President of Research and Information, United States Brewers Association. Personal interview.

Kelsey, R., 1970, "World's Fastest Bottling Lines," *Modern Packaging*, vol. 43, no. 4, pp. 138–139.

Kinnary, J. W., 1965, "Aerosol Workhorse: The Tinplate Can," *Drug and Cosmetic Industry*, vol. 97, no. 1, pp. 51+.

Kopp, R. J., and V. K. Smith, 1980, "Measuring Factor Substitution with Neoclassical Models: An Experimental Evaluation," *Bell Journal of Economics*, vol. 11, pp. 631–655.

Lead Development Association, 1964, *Lead Based Solders: A Survey of Their Use in New Motor Cars in the U.K.*, London.

Leimgruber, V. P., 1978, "The Welded Seam in Tinplate Fabrication," paper presented at Australian Seminar, International Tin Research Council, London.

Lueck, R. H., 1942, "Metal Container Changes in the Interest of Tin Conservation," *Proceedings: Annual Meeting*, Institute of Food Technology, Minneapolis, pp. 128–139.

———, and K. W. Brighton, 1944, "Metallic Substitutes for Hot-Dipped Tinplate," *Industrial and Engineering Chemistry*, vol. 36, no. 6, pp. 532–540.

Mason, J. Philip, and Joseph F. Manning, 1945, *The Technology of Plastics and Resins*, New York: Van Nostrand.

McKie, J., 1959, *Tin Cans and Tinplate*, Cambridge, Mass.: Harvard University Press.

McMahon, Albert D., 1965, *Copper: A Materials Survey*, Bureau of Mines, Washington, D.C.: Government Printing Office.

Medich, Joseph J., 1979, Marketing Services, Brockway Glass Company. Personal interview.

Melville, D. F., 1976, "Metal Can: State of the Art in the United States," paper presented to the First International Tin Conference, London.

Metallgesellschaft, annual, *Metal Statistics*, Frankfurt am Main, Federal Republic of Germany.

Modern Metals, monthly, Modern Metals Publishing, Chicago.

Modern Packaging, monthly, Breskin and Charlton Publishing, New York.

Modern Plastics, monthly, Society of the Plastics Industry, New York.

Modern Plastics Encyclopedia, 1975–1976, vol. 52, no. 10A, New York: McGraw-Hill.

Morley, R. A., and T. L. Wilkinson, 1978, "The Repair of Aluminum Heat Exchangers," *Welding Journal*, vol. 57, no. 10, pp. 15–20.

Motor Vehicle Manufacturers Association of the United States, Inc., 1979, *Facts and Figures*, Detroit.

———, 1979, *Information Handbook*, Detroit.

Naitove, Matthew H., Managing Editor, Plastics Technology. Personal interview.

National Academy of Sciences, 1941, Advisory Committee on Metals and Materials, *Report on Tin*, Washington, D.C.

National Canners Association, various years, *Canned Food Pack Statistics*, Washington, D.C.

National Materials Advisory Board, 1970, *Trends in the Use of Tin*, Washington, D.C.

National Materials Advisory Board, 1972, *Mutual Substitutability of Aluminum and Copper*, Washington, D.C.

National Sanitation Foundation, 1955, *A Study of Plastic Pipe for Potable Water Supplies*, University of Michigan, Ann Arbor.

National Soft Drink Association, 1976–1978, *Statistical Profile: The Soft Drink Industry of the United States*, Washington, D.C.

Parry, L., 1905, "Tin," *Mineral Resources of the United States Yearbook*, U.S. Department of the Interior, Washington, D.C.

Patton, Richard W., 1979, Corporate Planning, Aluminum Company of America. Personal interview.

Placey, R. J., 1978, "Setting the Record Straight on the Outlook for the Draw and Iron Beverage Can," paper presented at the Modern Can Manufacturing Clinic of the Society of Manufacturing Engineers, Oakland.

Plastic Pipe Institute, annual, *Statistics*, New York.

Plastics Focus, weekly, DWV Materials Authorized by Local Codes, vol. 12, 1980, New York.

Polysonics, 1970, *Noise and Vibration Characteristics of Soil Pipe Systems*, Washington, D.C.

Pratt, H. C., 1952, "Tin in Automobile Body Construction," *Symposium on Tin*, Special Publication No. 141, American Society for Testing Materials, Philadelphia, pp. 52–56.

Reinsch, S. P., 1979, *Food in Lead Soldered Cans: An Economic Analysis*, U.S. Food and Drug Administration, Washington, D.C.

Richardson, G. D., 1978, "Tinplate vs. Aluminum Two Piece Can Technology," unpublished paper, Metal Container Group, Ball Corporation, Australia.

Ricker, Chester S., 1948, "Ford's Radical Radiator Production Line," *American Machinist*, vol. 92, no. 14, pp. 77–81.

Rosenberg, N., 1973, "Innovative Responses to Materials Shortages," *American Economic Review*, vol. 63, no. 2, pp. 111–118.

Sanders, William T., 1979, Tin Mill Products, National Steel Corporation. Personal interview.

Sarvetnick, Harold A., 1969, *Polyvinyl Chloride*. New York: Van Nostrand-Reinhold.

Schumpeter, J. A., 1950, *Capitalism, Socialism and Democracy*, 3rd ed., New York: Harper & Row.

Seehafer, Marlyn E., 1979, Manager-Market Development, United States Steel Corporation. Personal interview.

Sellinger, Frank J., 1974, Statement Before the Panel on the Materials Policy of the Subcommittee of Environmental Pollution of the Committee of Public Works, United States Senate (July 11, 1974).

Shea, W. P., 1950, "Tin Industry-1950," unpublished paper, C. Tennant Sons, New York.

Silkotch, Mitchel S., Jr., Field Service Engineer-Thermoplastics, Interstab Chemicals. Personal interview.

Skinner, B. J., 1976, "A Second Iron Age Ahead?," *American Scientist*, vol. 64, pp. 258–269.

Slade, M. E., 1981, "Recent Advances in Econometric Estimation of Materials Substitution," *Resources Policy*, vol. 7, pp. 103–109.

Sosnin, H. A., 1974, "Choosing a Filler Metal for Copper Water Tube Joints," *Welding Journal*, vol. 53, no. 3, pp. 150–184.

Stanford Research Institute, 1975, *Chemical Economics Handbook*, Menlo Park, Calif.

Steel, 1946, "Fabricating Aluminum Radiators," vol. 119, no. 25, p. 102.

Steel, E. W., and T. J. McGhee, 1979, *Water Supply and Sewerage*, 5th ed., New York: McGraw-Hill.

Storck, W., 1979, "ABS Market a Mix of Optimism and Problems," *Chemical & Engineering News*, vol. 57, no. 31, pp. 8–9.

Tin, 1952, "Tinplate/Tin Printing/Canning," vol. 25, no. 1, p. 13.

———, 1954, "Tinplate/Tin Printing/Canning," vol. 27, no. 1, p. 4.

Tin and Its Uses, 1940, "Solder for Automobile Bodies," no. 7, pp. 13–14.

Tin International, 1966, "Patent for Canco's HTF Can," vol. 39, no. 10, p. 270.

———, 1977, "Tinplate/Tin Printing/Canning," vol. 50, no. 4, p. 3.

———, monthly, Tin Publications, London.

Tin Investigation, 1934, Report of the Subcommittee of the House Committee on Foreign Affairs, Parts I–III, Washington, D.C.

United Nations, 1969, *Fibro-Cement Composites*, Report and Proceedings of Expert Working Group Meeting, Vienna.

U.S. Brewers Association, 1950–1977, *Brewers Almanac*, Washington, D.C.

———, 1977, *Reference Source Book*, Washington, D.C.

U.S. Bureau of Census, annual, *Current Industrial Reports*, Washington, D.C.

U.S. Bureau of Mines, 1968, *Dictionary of Mining, Mineral and Related Terms*, Washington, D.C.

———, 1978, *Mineral Commodity Summaries*, Washington, D.C.

———, various years, *Minerals Yearbook*, Washington, D.C.

U.S. Council of Economic Advisers, annual, *Economic Report of the President*, Washington, D.C.

U.S. International Trade Commission, annual, *Synthetic Organic Chemicals, U.S. Production and Sales*, Washington, D.C.

U.S. National Bureau of Standards Building Science Series 111, 1978, *Investigation of Standards, Performance Characteristics and Evaluation Criteria for Thermoplastic Piping in Residential Plumbing Systems*, Washington, D.C.: U.S. Government Printing Office.

U.S., The President's Materials Policy Commission, 1952, *The Promise of Technology*, vol. IV of *Resources for the Future*, Washington, D.C.

U.S., War Metallurgy Committee, 1943, "Safe Substitute Solder for Food Cans," *Metal Progress*, vol. 44, no. 3, pp. 420 + .

Van Vleet, H. S., 1948, "Engineering the Tin Can," *Mechanical Engineering*, vol. 70, no. 4, pp. 315–320.

Way, Walter D., 1979, Can Manufacturers Institute Chairman of Statistics, Continental Can Company. Personal interview.

Weinberg, R. S., 1971, "The Effects of Convenience Packaging on the Malt Beverage Industry 1947–1969," *Reference Source Book*, Washington, D. C.: U.S. Brewer's Association, 40 pp.

———, 1978, "The Economics of the Malt Beverage Production/ Distribution System 1976," unpublished paper, R. S. Weinberg and Associates, St. Louis.

———, 1979, President, R. S. Weinberg and Associates. Personal interview.

Woodroof, J., and G. Phillips, 1974, *Beverages: Carbonated and Noncarbonated*, Westport, Conn.: AVI Publishing.

Woods, D. W., and J. C. Burrows, 1980, *The World Aluminum-Bauxite Market*, New York: Praeger, pp. 184–230.

Index

About the Authors

John E. Tilton is professor of mineral economics at The Pennsylvania State University and co-director with Hans Landsberg of the Mineral Economics and Policy Program at Resources for the Future. He is the co-author of *Public Policy and the Diffusion of Technology* (Pennsylvania State University Press, 1978) and author of *The Future of Nonfuel Minerals* (Brookings Institution, 1978). Currently, he is director of the Minerals Trade and Markets Project at the International Institute for Applied Systems Analysis, Laxenburg, Austria.

Patrick D. Canavan is principal research officer for policy and planning in the department of minerals, Papua New Guinea. Frederick R. Demler is a commodity analyst and Derek G. Gill is a senior planning analyst with the Exxon Minerals Company.